ENVIRONMENTAL SCIENCE, ENGINEERING AND TECHNOLOGY

PESTICIDE USE IN THE U.S.: LAW AND FEE ISSUES

ENVIRONMENTAL SCIENCE, ENGINEERING AND TECHNOLOGY

Additional books in this series can be found on Nova's website under the Series tab.

Additional E-books in this series can be found on Nova's website under the E-book tab.

ENVIRONMENTAL SCIENCE, ENGINEERING AND TECHNOLOGY

PESTICIDE USE IN THE U.S.: LAW AND FEE ISSUES

REGINA M. BIANCHI
EDITOR

Nova Science Publishers, Inc.
New York

For permission to use material from this book please contact us:
Telephone 631-231-7269; Fax 631-231-8175
Web Site: http://www.novapublishers.com

LIBRARY OF CONGRESS CATALOGING-IN-PUBLICATION DATA

Pesticide use in the U.S. : law and fee issues / editors, Regina M. Bianchi.
p. cm.
Includes index.
ISBN 978-1-61122-053-7 (hardcover)
1. Pesticides--Law and legislation--United States. 2.
Pesticides--Environmental aspects--United States. 3. Pesticide residues in
food--Law and legislation--United States. 4. United States. Pesticide
Registration Improvement Act. I. Bianchi, Regina M.
KF3959.P4735 2010
344.7304'6334--dc22

2010035907

Published by Nova Science Publishers, Inc. † New York

CONTENTS

PREFACE

The Federal Insecticide, Fungicide, and Rodenticide Act (FIFRA) directs the EPA to restrict the use of pesticides as necessary to prevent unreasonable adverse effects on people and the environment, taking into account the costs and benefits of various pesticide use. There are an estimated 18,000 pesticide products currently in use. These generally are regulated under the FIFRA. This book provides a brief history of the federal pesticide laws and describes key provisions of the laws; including the pesticide registration process and how it interfaces with food safety requirements.

Chapter 1- This report summarizes the major statutory authorities governing pesticide regulation: the Federal Insecticide, Fungicide, and Rodenticide Act (FIFRA), and Section 408 of the Federal Food, Drug, and Cosmetic Act (FFDCA), as well as the major regulatory programs for pesticides. Text relevant to FIFRA is excerpted, with minor modifications, from the corresponding chapter of CRS Report RL30798, *Environmental Laws: Summaries of Major Statutes Administered by the EPA*, coordinated by Bonnie C. Gitlin, which summarizes more than a dozen environmental statutes.

Congress enacted the original version of FIFRA in 1947, but a revision in 1972 is the basis of current pesticide policy. Substantial changes were made in 1988 and again in 1996. The 1996 FIFRA amendments were contained in the Food Quality Protection Act (FQPA), which also amended the FFDCA. Congress first required limits on pesticide residues on raw food in 1954 amendments to the FFDCA. Limits were required for food additives (including pesticide residues in processed foods) in the 1958 FFDCA amendments. In the 1996 FFDCA amendments, Congress established a new standard of safety for pesticide residues in food (both raw and processed): maximum residue levels set by EPA must ensure with "a reasonable certainty" that "no harm" will result from pesticide exposure.

Chapter 2- This report provides background on the emerging conflict over interpretation and implementation of the Federal Insecticide, Fungicide, and Rodenticide Act (FIFRA) and the Clean Water Act (CWA). For the more than 30 years since they were enacted, there had been little apparent conflict between them. But their relationship has recently been challenged in several arenas, including the federal courts and regulatory proceedings of the Environmental Protection Agency (EPA). In this report, a brief discussion of the two laws is followed by a review of the major litigation of interest. EPA's efforts to clarify its policy in this area are discussed, including a regulation issued in 2006 that was subsequently vacated by a federal court, as well as possible options for EPA and Congress to address the issues further.

FIFRA governs the labeling, distribution, sale, and use of pesticides, including insecticides and herbicides. Its objective is to protect human health and the environment from unreasonable adverse effects of pesticides. It establishes a nationally uniform labeling system requiring the registration of all pesticides sold in the United States, and requiring users to comply with the national label. The CWA creates a comprehensive regulatory scheme to control the discharge of pollutants into the nation's waters; the discharge of pollutants without a permit violates the act.

Chapter 3- The U.S. Environmental Protection Agency (EPA) is responsible for regulating the sale, use, and distribution of pesticides under the authority of two statutes. The Federal Insecticide, Fungicide, and Rodenticide Act (FIFRA) (7 U.S.C. 136-136y), a licensing statute, requires EPA to review and register the use of pesticide products. The Federal Food, Drug, and Cosmetic Act (FFDCA) (21 U.S.C. 346a) requires the establishment of maximum limits (tolerances) for pesticide residues on food in interstate commerce. EPA was also required to reevaluate older, registered pesticides (i.e., "reregistration" for pesticides registered prior to 1984, and more recently, registration review) and to reassess existing tolerances (i.e., tolerance reassessment) to ensure they meet current safety standards. Although U.S. Treasury revenues cover most costs for administering these acts, fees paid by pesticide manufacturers and other registrants have supplemented EPA appropriations for many years as a means of increasing the pace of the agency's activities under FIFRA and FFDCA.

Chapter 4- The Consolidated Appropriations Act of 2004 established a new system for registering pesticides, called the Pesticide Registration Improvement Act, or PRIA. The new section 33 of the Federal Insecticide,

Fungicide, and Rodenticide Act (FIFRA), PRIA created a registration service fee system for applications for specified pesticide registration, amended registration, and associated tolerance actions, which set maximum residue levels for food and feed. Under PRIA, fees are charged for covered applications received on or after March 23, 2004, and for certain pending applications received before that date. The Environmental Protection Agency (EPA) is required to make a determination on the application within the decision times specified. The fee system was authorized until September 30, 2010. Due to the efforts of the PRIA Coalition of industry, trade associations, and public interest groups, PRIA was reauthorized on October 9, 2007 effective retroactively to October 1, 2007, the beginning of Fiscal Year 2008. The Pesticide Registration Improvement Renewal Act (PRIA 2) authorized the fee system to September 30, 2012.

In: Pesticide Use in the U.S.: Law and Fee Issues ISBN: 978-1-61122-053-7
Editors: Regina M. Bianchi

Chapter 1

PESTICIDE LAW: A SUMMARY OF THE STATUTES

Linda-Jo Schierow

SUMMARY

This report summarizes the major statutory authorities governing pesticide regulation: the Federal Insecticide, Fungicide, and Rodenticide Act (FIFRA), and Section 408 of the Federal Food, Drug, and Cosmetic Act (FFDCA), as well as the major regulatory programs for pesticides. Text relevant to FIFRA is excerpted, with minor modifications, from the corresponding chapter of CRS Report RL30798, *Environmental Laws: Summaries of Major Statutes Administered by the EPA*, coordinated by Bonnie C. Gitlin, which summarizes more than a dozen environmental statutes.

Congress enacted the original version of FIFRA in 1947, but a revision in 1972 is the basis of current pesticide policy. Substantial changes were made in 1988 and again in 1996. The 1996 FIFRA amendments were contained in the Food Quality Protection Act (FQPA), which also amended the FFDCA. Congress first required limits on pesticide residues on raw food in 1954 amendments to the FFDCA. Limits were required for food additives (including pesticide residues in processed foods) in the 1958 FFDCA amendments. In the 1996 FFDCA amendments, Congress established a new standard of safety for pesticide residues in food (both raw and processed): maximum residue levels set by EPA must ensure with "a reasonable certainty" that "no harm" will result from pesticide exposure.

FIFRA requires the U.S. Environmental Protection Agency (EPA) to regulate the sale and use of pesticides in the United States through registration and labeling of pesticide products. The sale of any pesticide is prohibited in the United States unless it is registered and labeled. EPA is directed to restrict the use of pesticides as necessary to prevent unreasonable adverse effects on people and the environment, taking into account the costs and benefits of various pesticide uses. In addition, FIFRA requires EPA to reregister older pesticides based on new data that meet current regulatory and scientific standards. Pesticides manufactured solely for export do not require registration. For pesticides to be registered for use in food production, FFDCA Section 408 authorizes EPA to establish allowable residue levels, called "tolerances," that ensure that human exposure to pesticide residues in food will be "safe." Foods with pesticide residues above the tolerance, or for which there is no tolerance established, may not be imported or sold in interstate commerce. A pesticide may not be registered under FIFRA for a food use unless a tolerance for that pesticide and food has been established under FFDCA.

FIFRA directs EPA to make public any data submitted to support a registration application, if EPA registers the pesticide, but certain data are protected as trade secrets, and other registrants may not use the same data to support registration applications for similar pesticides for a period of 10 years. EPA continues to evaluate the safety of pesticides after they are registered, as new information becomes available. A pesticide registration may be canceled or amended if EPA determines that current use may cause unreasonable adverse effects.

INTRODUCTION

The Environmental Protection Agency (EPA) is responsible for implementing federal pesticide policies under two statutes: the Federal Insecticide, Fungicide, and Rodenticide Act (FIFRA),[1] governing the sale and use of pesticide products within the United States, and the Federal Food, Drug, and Cosmetic Act (FFDCA), which limits pesticide residues on food in interstate commerce (including imports). This report defines key terms, provides a brief history of the federal pesticide laws, and describes key provisions of the laws, including the pesticide registration process and how it interfaces with food safety requirements. In addition, this report lists several references for more detailed information about the acts, and two tables cross reference sections of the *U.S. Code* with corresponding sections of the acts. The report is descriptive rather than analytic, highlights key provisions rather than providing a comprehensive inventory of the acts' numerous sections, and addresses authorities and limitations imposed by statute, rather than the status of EPA implementation or other policy issues. Other CRS products address current pesticide issues, including CRS Report RL32218, *Pesticide Registration and Tolerance Fees: An Overview*, by Robert Esworthy, and CRS Report RL32884, *Pesticide Use and Water Quality: Are the Laws Complementary or in Conflict?*, by Claudia Copeland.

OVERVIEW

There are an estimated 18,000 pesticide products currently in use.[2] These generally are regulated under FIFRA, but approximately 5,800 pesticide products used in food production also are regulated under the FFDCA, as discussed below. FIFRA requires EPA to regulate the sale and use of pesticides in the United States through registration and labeling.[3] Pesticides are broadly defined in FIFRA Section 2(u) as chemicals and other products used to kill, repel, or control pests. Familiar examples include pesticides used to kill insects and weeds that can reduce the yield, and sometimes harm the quality, of agricultural crops, ornamental plants, forests, wooden structures, and pastures. But the broad definition of "pesticide" in FIFRA also applies to products with less familiar "pesticidal uses." For example, substances used to control mold, mildew, algae, and other nuisance growths on equipment, in surface water, or on stored grains are pesticides. The term also applies to disinfectants and sterilizing agents, animal repellents, rat poison, and many other substances.

FIFRA directs EPA to restrict the use of pesticides as necessary to prevent unreasonable adverse effects on people and the environment, taking into account the costs and benefits of various pesticide uses. The act prohibits sale of any pesticide in the United States unless it is registered and labeled to indicate approved uses and restrictions. It is a violation of the law to use a pesticide in a manner that is inconsistent with the label instructions. EPA registers each pesticide product for each approved use. For example, a product may be registered for use on green beans to control mites, as a seed treatment for cotton, and as a treatment for structural cracks. In addition, FIFRA requires EPA to reregister older pesticides based on new data that meet current regulatory and scientific standards. Establishments that manufacture or sell pesticide products must register with EPA. Facility managers are required to keep certain records and to allow inspections by federal or state regulatory officials.

For the 600 or more pesticides (i.e., active ingredients) registered for use in food production, the FFDCA Section 408 authorizes EPA to establish maximum allowable residue levels (called tolerances) that ensure that human exposure to the pesticide ingredients in food and animal feed will be "safe."[4] A "safe" tolerance is defined as a level at which there is "a reasonable certainty of no harm" from the exposure. Under FFDCA, foods with a residue of a pesticide ingredient for which there is no tolerance established, or with a residue level exceeding an established tolerance limit, are declared "unsafe" and "adulterated"; such foods cannot be sold in interstate commerce or imported to the United States. Pesticides may not be registered under FIFRA for use on food unless tolerances (or exemptions) have been established under the FFDCA.

HISTORY OF FEDERAL PESTICIDE LAW

Table 1 and **Table 2** summarize the history of FIFRA and FFDCA, respectively.

FIFRA

Federal pesticide legislation was first enacted in 1910. It aimed to reduce economic exploitation of farmers by manufacturers and distributors of adulterated or ineffective pesticides. Congress did not address the potential risks to human health posed by pesticide products until it enacted the original 1947 version of FIFRA. The U.S. Department of Agriculture (USDA) was responsible for administering the pesticide statutes during this period. However, responsibility was shifted to the EPA when that Agency was created in 1970. Broader congressional concerns about long- and short-term toxic effects of pesticide exposure on people who applied pesticides (applicators), wildlife, nontarget insects and birds, and on food consumers, subsequently led to a complete revision of FIFRA in 1972. The 1972 law completely replaced the original 1947 law, and is the basis of current federal policy. Substantial changes were made in 1988 (P.L. 100-532), 1996 (P.L. 104-170), and 2004 (P.L. 108-199). The 1988 amendments focused on accelerating the reregistration process. The 1996 amendments facilitated registration of pesticides for special (so- called "minor") uses, reauthorized collection of fees to support reregistration, and required coordination of regulations implementing FIFRA and the Federal Food, Drug, and Cosmetic Act (21 U.S.C. 321 *et seq.).* The 2004 amendments, known as the Pesticide Registration Improvement Act (PRIA), modified the types and amounts of fees that EPA could collect to support its activities. The Pesticide Registration Improvement Renewal Act, or PRIA 2 (P.L. 110-94), enacted October 9, 2007, reauthorized and revised these fee provisions, which would have expired at the end of FY2008. See Table 3 for a listing of current provisions in FIFRA.

Authorization for appropriations for FIFRA expired on September 31, 1991, although appropriations bills have continued to provide funding to implement the law. Authority provided by FIFRA to EPA to issue and enforce regulations is, for the most part, permanent, and is not affected by the lack of authorization.

FFDCA

The original Federal Food, Drug, and Cosmetic Act of 1938 (FFDCA) established the structure of the current law. With respect to food safety and possible contaminants, it required the Food and Drug Administration (then a part of the U.S. Department of Agriculture) to set maximum residue levels (tolerances) for unavoidable poisonous substances in food. Congress acted to protect consumers from pesticide residues on food in 1954 by adding a new Section 408 to the FFDCA. It directed FDA to set residue tolerances for all pesticides in *raw* agricultural commodities. Congress expanded the requirement for tolerances in the Food Additives Amendment of 1958, which added Section 409, directing FDA to set tolerances for food additives, including pesticide residues in *processed* foods. Section 409 also forbade the addition to food of any additive (including pesticide residue), if it was found to be a potential cancer-causing agent. This provision is referred to as the Delaney Clause.

Table 1. Federal Insecticide, Fungicide, and Rodenticide Act and Amendments (Codified Generally as 7 U.S.C. 136-136y)

Year	Act	Public Law Number
1947	Federal Insecticide, Fungicide, and Rodenticide Act	P.L. 80-104
1964	Federal Insecticide, Fungicide, and Rodenticide Act Amendments	P.L. 88-305
1972	Federal Environmental Pesticide Control Act	P.L. 92-516
1975	Federal Insecticide, Fungicide, and Rodenticide Act Extension	P.L. 94-140
1978	Federal Pesticide Act of 1978	P.L. 95-396
1980	Federal Insecticide, Fungicide and Rodenticide Act Amendments	P.L. 96-539
1988	Federal Insecticide, Fungicide, and Rodenticide Amendments of 1988	P.L. 100-532
1990	Food, Agriculture, Conservation, and Trade Act of 1990	P.L. 101-624
1991	Food, Agriculture, Conservation and Trade Amendments of 1991	P.L. 102-237
1996	Food Quality Protection Act (FQPA) of 1996	P.L. 104-170
2004	Pesticide Registration Improvement Act of 2003	P.L. 108-199
2007	Pesticide Registration Improvement Renewal Act	P.L. 110-94

Source: Congressional Research Service.

Note: The current FIFRA statute was established by P.L. 92-516, which completely replaced (by amendment) the original 1947 legislation.

In 1970, authority to establish tolerances for pesticide residues was transferred to the newly formed EPA. FDA (now in the Department of Health and Human Services) retained responsibility for enforcement of tolerances in food that is imported or sold across state boundaries.

In 1996, Congress substantially revised requirements for pesticide residue tolerance setting in the Food Quality Protection Act (FQPA). The FQPA redefined terms so that pesticide residues in processed foods were no longer regulated as food additives, and therefore no longer were subject to the Delaney Clause. The FQPA also established a new safety standard of a "reasonable certainty of no harm" from exposure to pesticides. See Table 4 for a listing of current pesticide- related provisions in FFDCA.

The Act of July 22, 1954, authorized such sums as may be necessary to carry out this FFDCA section (21 U.S.C. 346b).

Table 2. Federal Food, Drug, and Cosmetic Act, Section 408, and Amendments (Codified Generally as 21 USC 346a)

Year	Act	Public Law Number
1938	Federal Food, Drug, and Cosmetic Act	Act of June 25, 1938
1954	Federal Food, Drug, and Cosmetic Act Amendments	Act of July 22, 1954
1958	Food Additive Amendments of 1958 (including the Delaney Clause)	P.L. 85-929
1996	Food Quality Protection Act of 1996	P.L. 104-170

Source: Congressional Research Service.

REGISTRATION OF PESTICIDE PRODUCTS

When pesticide manufacturers apply to register a pesticide active ingredient, pesticide product, or a new use of a registered pesticide under FIFRA Section 3, EPA requires them to submit scientific data on toxicity and behavior in the environment. EPA may require data from any combination of more than 100 different tests, depending on the potential toxicity of active and inert ingredients and degree of exposure. To register a pesticide use on food, EPA also requires applicants to identify analytical methods that can be used to test food for residues of active ingredients, certain inert ingredients, and their breakdown products and to determine the amount of residue that could remain on crops, as well as on (or in) food products, assuming that the pesticide product is applied according to the manufacturers' recommended rates and methods.

Based on the data submitted, EPA determines whether and under what conditions the proposed pesticide use would present an unreasonable risk to human health or the environment. If the pesticide is proposed for use on a food crop, EPA also determines whether a "safe" level of pesticide residue, called a "tolerance," can be established under the Federal Food, Drug, and Cosmetic Act. A tolerance must be established before a pesticide registration may be granted for use on food crops. If registration is granted, the Agency specifies the approved uses and conditions of use, including safe methods of pesticide storage and disposal, which the registrant must explain on the product label. FIFRA requires that federal regulations for pesticide labels preempt state, local, and tribal regulations. Use of a pesticide product in a manner inconsistent with its label is prohibited.

EPA may classify and register a pesticide product for general or for restricted use. Products known as "restricted-use pesticides" are those judged to be more dangerous to the applicator or to the environment. Such pesticides can be applied only by people who have been trained and certified. Individual states and Indian tribes generally are responsible for training and certifying pesticide applicators.

FIFRA Section 3 also allows "conditional," temporary registrations if (1) the proposed pesticide ingredients and uses are substantially similar to currently registered products and will not create additional significant environmental risks; (2) an amendment is proposed for additional uses of a registered pesticide, and sufficient data are submitted indicating that there is no significant additional risk; or (3) data requirements for a new active ingredient require more time to generate than normally allowed, and use of the pesticide during the period will not cause any unreasonable adverse effect on the environment and will be in the public interest.

FIFRA-FFDCA Coordination

EPA has long coordinated pesticide registrations for food uses under FIFRA with tolerance setting under the FFDCA. The Food Quality Protection Act of 1996 (FQPA; P.L. 104-170) codified this policy. Thus, if EPA revokes a residue tolerance under FFDCA, it cancels the FIFRA pesticide registration for that food use. Similarly, if a pesticide registration for use on a food crop is canceled, EPA also cancels the residue tolerance for the food. However, just as FIFRA allows continued use of remaining pesticide stocks after a registration is canceled, FFDCA allows continued commerce in commodities legally treated with a pesticide. Thus, EPA does not immediately revoke the tolerance for the pesticide residue, when it cancels the corresponding registration.

Tolerance Setting

Any person who has registered a pesticide may petition EPA proposing establishment of a tolerance or an exemption for that pesticide to permit its use on food-related crops.[5] Tolerance petitions must include information about pesticide application rates, measured concentrations of pesticide residues on the food after the pesticide has been applied according to directions on its label, and safety of pesticide use on food crops. The FFDCA requires EPA to respond to each petition by establishing a tolerance or exempting the pesticide from the requirement. If the pesticide will not leave residues above an established safe level, EPA will register the pesticide for use on that food product and set the tolerance level by issuing a regulation. EPA tolerances for pesticide residues preempt state and local restrictions on food, if the state and local restrictions are based on lower residue levels. States may petition for an exception if the EPA-set residue level threatens public health.

The FFDCA, Section 408, as amended, requires EPA to assess safety in terms of total exposure to the pesticide (that is, to the concentration of pesticide allowed by the tolerance, together with all other dietary and non-food exposures for which there is reliable information) as well as to other pesticides that have the same toxic effects on people. No quantitative standard of safety is established by law, but the Committee on Commerce noted in its report on the bill that became the FQPA that EPA should continue setting standards to ensure safety as it had in the past:

> ... the Committee expects that a tolerance will provide a 'reasonable certainty of no harm' if the Administrator determines that the aggregate exposure to the pesticide chemical residue will be lower by an ample margin of safety than the level at which the pesticide chemical residue will not cause or contribute to any known or anticipated harm to human health. The Committee further expects, based on discussions with the Environmental Protection Agency, that the Administrator will interpret an ample margin of safety to be a 100-fold safety factor applied to the scientifically determined 'no observable effect' level when data are extrapolated from animal studies.[6]

In determining a safe level, the FFDCA directs EPA to take into account many factors, including available information on dietary exposure to pesticides among infants and children. FQPA strictly limited the nature and influence of benefits considered in tolerance setting under Section 408 of the FFDCA. As amended, Section 408 allows EPA to maintain or modify existing tolerances (but not to establish new tolerances) at higher than "safe" residue levels *only if* the pesticide use avoids other greater risks to consumers, or is necessary to avoid significant disruption in domestic production of an adequate, wholesome, and economical food supply. Such higher tolerance levels may be set only for pesticides that are potential carcinogens (or have some other health effect) for which there is no known level of exposure at which no harm is anticipated (known as a non-threshold effect). The higher tolerance level allowed for such pesticide residues must be "safe" for infants and children, as well as with respect to health effects for which there is a known threshold (that is, a level below which exposure is known to be harmless). The higher cancer (or other non-threshold) risk posed by the tolerance on an annual basis may not be more than 10 times the risk at a "safe" level of exposure and not more than twice the risk of a "safe" level over a lifetime.

For nonthreshold effects, the House Commerce Committee provided additional guidance for establishing a level of residue that should be considered "safe."

> In the case of a nonthreshold effect which can be assessed through quantitative risk assessment, such as a cancer effect, the Committee expects, based on its understanding of current EPA practice, that a tolerance will be considered to provide a 'reasonable certainty of no harm' if any increase in lifetime risk, based on quantitative risk assessment using conservative assumptions, will be no greater than 'negligible.' It is the Committee's understanding that, under current EPA practice, ... EPA interprets a negligible risk to be a one-in-a-million lifetime risk. The Committee expects the Administrator to continue to follow this interpretation.[7]

The "safe" standard applies to both raw and processed foods, and requires EPA to consider cumulative and aggregate exposure to pesticides in food, drinking water, air, and consumer products. Congress directed EPA to reevaluate all existing tolerances against this standard before August 2006.

FFDCA directs the FDA in the Department of Health and Human Services and USDA to monitor pesticide residue levels in food in interstate commerce and to enforce tolerances through their food inspection programs. USDA is responsible for inspecting meat and poultry; FDA inspects all other foods. States also may monitor pesticide residues in food sold within their jurisdictions.

Public Disclosure, Exclusive Use, and Trade Secrets

FIFRA Section 3 directs EPA to make the data submitted by the applicant for pesticide registration publicly available within 30 days after a registration is granted. However, applicants may claim certain data are protected as trade secrets under FIFRA, Section 10. If EPA agrees that the data are protected, the Agency must withhold those data from the public, unless the data pertain to the health effects or environmental fate or effects of the pesticide ingredients. Information may be protected if it qualifies as a trade secret and reveals: (1) manufacturing processes; (2) details of methods for testing, detecting, or measuring amounts of inert ingredients; or (3) the identity or percentage quantity of inert ingredients.

Companies sometimes seek to register a product based upon the registration of similar products, relying upon the data provided by the original registrant that are publicly released. This is allowed. However, Section 3 of FIFRA provides for a 10-year period of "exclusive use" by the registrant of data submitted in support of an original registration or a new use. In addition, an applicant who submits any new data in support of a registration is entitled to compensation for the cost of data development by any subsequent applicant who supports an application with that data within 15 years of its submission. If compensation is not jointly agreed upon by the registrant and applicant, binding arbitration can be invoked.

Reregistration

Most pesticides currently registered in the United States are older pesticides and were not subject to modern safety reviews. Amendments to FIFRA in 1972 directed EPA to "reregister" approximately 35,000 older products, in order to assess their safety in light of current standards. The task of reregistering older pesticides was streamlined by reviewing groupings of products having the same active ingredients, on a generic instead of individual product basis. For food-use pesticides, EPA evaluated a pesticide's eligibility for reregistration at the same time the Agency reassessed the tolerance for that pesticide under the FFDCA. The FQPA required EPA to reassess pesticides posing the greatest risks first. Many of the 35,000 pesticide products were not reviewed and their registrations were canceled, because registrants did not request reregistration. At least 14,000 products are no longer in use. Nevertheless, the task for registrants and EPA was immense and costly.

To accelerate the process of reregistration, Congress, in 1988 amendments to FIFRA, imposed a 10-year reregistration schedule. To help pay for the additional costs of the accelerated process, Congress directed EPA to require registrants to pay reregistration and annual registration maintenance fees on pesticide ingredients and products. The 1996 amendments to FIFRA extended EPA's authority to collect maintenance fees through FY2001. Exemptions from, or reductions in, fees were allowed for minor-use pesticides, public health pesticides, and small business registrants. Congress extended authority for fees annually through appropriations legislation after FY2001, until the omnibus appropriations legislation signed January 23, 2004 (P.L. 108-199) modified the types and amounts of fees that EPA could collect, potentially through FY2008.

The 2004 FIFRA amendments (PRIA) reauthorized collection of annual "maintenance" fees to support registration, designated a portion of those fees for the review of inert ingredients, and extended the deadline for completion of reregistration. PRIA directed EPA to complete Reregistration Eligibility Decisions (REDs) for pesticides with food uses/tolerances by August 3, 2006, and to complete REDs for all remaining non-food use pesticides by October 3, 2008. The reregistration process will continue for several years after that date, as explained on the EPA reregistration website:

> After EPA has issued a RED and declared a pesticide eligible for reregistration, individual end-use products that contain the pesticide active ingredient still must be reregistered. Through this concluding part of the process, known as "product reregistration," the Agency makes sure that the risk reduction measures called for in REDs are reflected on individual pesticide product labels. In some cases, the Agency uses Memoranda of Agreement or other measures to include risk reduction measures on pesticide labels sooner, before product reregistration is completed. EPA plans to complete the last product reregistration decisions several years after the last REDs are signed.[8]

EPA authority for administering these fees would have expired at the end of FY2008, but it was extended by the Pesticide Registration Improvement Renewal Act, or PRIA 2 (P.L. 110-94), enacted October 9, 2007, effective retroactively to the beginning of FY2008 through FY20 12. PRIA 2 also made some technical revisions, primarily modifications to the fee payment process and an expansion of the range of categories of pesticide registration (licensing) activities subject to fees.

SPECIAL REVIEW

EPA continues to evaluate the safety of pesticides after they are registered as new information becomes available. FIFRA requires registrants to report promptly any new evidence of adverse effects from pesticide exposure. If evidence indicates that a registered pesticide may pose an unreasonable risk, EPA may initiate a special review of available information to reevaluate the risks and benefits of each registered use. FIFRA also authorizes EPA to require registrants to conduct new studies to fill gaps in scientific understanding to assist risk assessments. As a result of a special review EPA may conclude that registration is adequate, needs amendment, or should be canceled.

CANCELING OR SUSPENDING A REGISTRATION

If a special review or reregistration evaluation finds that a registered use may cause "unreasonable adverse effects," EPA may amend or cancel the registration.[9] FIFRA also allows registrants to request cancellation or amendment of a registration to terminate selected pesticide uses. Requesting voluntary cancellation sometimes reflects a registrant's conclusion that the cost of additional studies is not worth the expected benefit (that is, profit) from sales if the registration would be maintained.

If a registration is canceled for one or more uses of a pesticide, FIFRA does not permit it to be sold or distributed for those uses in the United States, although for a specified period of time, U.S. farmers may use remaining stocks, and commerce may continue for commodities that were legally treated with the pesticide. FIFRA allows registrants to appeal an EPA decision to cancel a registration. An appeal initiates a lengthy review process during which the product may continue to be marketed. However, if there is threat of an "imminent hazard" during the time required to cancel a registration, FIFRA authorizes EPA to suspend registration. Suspension orders, which also may be appealed, stop sales and use of the pesticide. In the event of suspension and cancellation, FIFRA Section 15 directs EPA to request an appropriation from Congress to compensate anyone who owned any of the pesticide and suffered any loss due to the suspension or cancellation. The registrant of the suspended and canceled product is responsible, however, for all of the transportation and disposal costs, and most storage costs.

USE OF UNREGISTERED PESTICIDES

FIFRA also allows for unregistered use of pesticide products in special circumstances. Section 5 allows experimental use permits for purposes of research and to collect data needed to register a pesticide. Section 18 allows "emergency exemptions" from the provisions of FIFRA to be granted to federal or state agencies, for example, if there is a virulent outbreak of a disease that cannot be controlled by registered products. In addition, Section 24(c) permits states to allow additional uses of a federally registered product to meet "special local needs."

ENFORCEMENT

Generally, EPA has the authority to enforce FIFRA requirements. However, FIFRA Section 26 gives states with adequate enforcement procedures, laws, and regulations primary authority, including inspection authority, for enforcing FIFRA provisions related to pesticide use. EPA is authorized by Section 27 to rescind a state's primary enforcement responsibility if it is not being carried out.

FIFRA Section 11 authorizes EPA to form cooperative agreements with states, giving them the responsibility for training and certifying applicators of restricted use pesticides. States also may initially review and give preliminary approval to applications for emergency exemptions and special local needs registrations (although under some conditions FIFRA allows EPA later to deny state-approved applications).

Section 9 authorizes inspections by EPA and authorized state officials of pesticide products where they are stored for distribution or sale. Section 13 authorizes EPA to issue orders to stop sales and to seize supplies of pesticide products. Civil and criminal penalties for violations of FIFRA are established in Section 14, while Section 15 provides indemnity payments for end users, distributors, and dealers of pesticides when registrations are suspended and canceled.

Federal district courts are authorized in Section 16 to review EPA final actions and omissions when action is not discretionary. People adversely affected by an EPA order may file for judicial review of the order following a hearing. But, FIFRA does not authorize citizen suits against violators.

EXPORT OF UNREGISTERED PESTICIDES

FIFRA does not give EPA the authority to regulate domestic production for export of unregistered pesticides, even if U.S. registration has been canceled for health or environmental reasons. However, FIFRA does require exporters to prepare or pack pesticides as specified by the purchaser and in accord with some of the FIFRA labeling provisions. For example, exporters must translate warning information into the language of the destination. FIFRA also requires exporters of unregistered pesticides to obtain the purchaser's signature on a statement acknowledging that the pesticide is unregistered and cannot be sold in the United States. EPA is required to notify governments of other countries and international agencies whenever a registration, cancellation, or suspension of any pesticide becomes or ceases to be effective in the United States.

Table 3. Major U.S. Code Sections of the Federal Insecticide, Fungicide, and Rodenticide Act (Codified Generally as 7 U.S.C. 136-1 36y)

7 U.S.C.	Section Title	FIFRA
	Short title and table of contents	sec. 1
136	Definitions	sec. 2
136a	Registration of pesticides	sec. 3
136a-1	Reregistration of registered pesticides	sec. 4
136c	Experimental use permits	sec. 5
136d	Administration review; suspension	sec. 6
136e	Registration of establishments	sec. 7
136f	Books and records	sec. 8
136g	Inspection of establishments	sec. 9
136h	Protection of trade secrets and other information	sec. 10
136i	Restricted use pesticides; applicators	sec. 11
136j	Unlawful acts	sec. 12
136k	Stop sale, use, removal, and seizure	sec. 13
136l	Penalties	sec. 14
136m	Indemnities	sec. 15
136n	Administrative procedure; judicial review	sec. 16
136o	Imports and exports	sec. 17
136p	Exemption of federal and state agencies	sec. 18

Table 3. (Continued)

7 U.S.C.	Section Title	FIFRA
136q	Storage, disposal, transportation, and recall	sec. 19
136r	Research and monitoring	sec. 20
136s	Solicitation of comments; notice of public hearings	sec. 21
136t	Delegation and cooperation	sec. 22
136u	State cooperation, aid, training	sec. 23
136v	Authority of states	sec. 24
136w	Authority of Administrator	sec. 25
136w-1	State primary enforcement responsibility	sec. 26
136w-2	Failure by the state to assure enforcement of state pesticide use regulations	sec. 27
136w-3	Identification of pests; cooperation with Department of Agriculture's program	sec. 28
136w-4	Annual report	sec. 29
136w-5	Minimum requirements for training of maintenance applicators and service technicians	sec. 30
136w-6	Environmental Protection Agency minor use program	sec. 31
136w-7	Department of Agriculture minor use program	sec. 32
136w-8	Pesticide Registration Service Fees	sec. 33
136x	Severability	sec. 34
136y	Authorization of Appropriations	sec. 35

Note: This table shows only the major code sections. For more detail and to determine when a section was added, the reader should consult the official printed version of the U.S. Code.

Table 4. Major U.S. Code Sections of the Federal Food, Drug, and Cosmetic Act Related to Pesticides (Codified Generally as 21 U.S.C. 321 - 346a)

21 U.S.C.	Section Title	FFDCA
Chapter II—Definitions		
321	Definitions	sec. 201
Chapter III—Prohibited Acts and Penalties		
331	Prohibited acts	sec. 301
332	Injunction proceedings	sec. 302
333	Penalties	sec. 303
334	Seizure	sec. 304

Table 4. (Continued)

21 U.S.C.	Section Title	FFDCA
Chapter IV—Food		
342	Adulterated food	sec. 402
343	Misbranded food	sec. 403
346	Tolerances for poisonous ingredients in food	sec. 406
346a	Tolerances and exemptions for pesticide chemical residues	sec. 408
346a(a)	Requirement for tolerance or exemption	sec. 408(a)
346a(b)	Authority and standard for tolerance	sec. 408(b)
346a(c)	Authority and standard for exemptions	sec. 408(c)
346a(d)	Petition for tolerance or exemption	sec. 408(d)
346a(e)	Action on Administrator's own initiative	sec. 408(e)
346a(f)	Special data requirements	sec. 408(f)
346a(g)	Effective data, objections, hearings, and administrative review	sec. 408(g)
346a(h)	Judicial review	sec. 408(h)
346a(i)	Confidentiality and use of data	sec. 408(i)
346a(j)	Status of previously issued regulations	sec. 408(j)
346a(k)	Transitional provision	sec. 408(k)
346a(l)	Harmonization with action under other laws	sec. 408(l)
346a(m)	Fees	sec. 408(m)
346a(n)	National uniformity of tolerances	sec. 408(n)
346a(o)	Consumer right to know	sec. 408(o)
346a(p)	Estrogenic substances screening program	sec. 408(p)
346a(q)	Schedule for review	sec. 408(q)
346a(r)	Temporary tolerance or exemption	sec. 408(r)
346a(s)	Savings clause	sec. 408(s)

Note: This table shows only the major code sections. For more detail and to determine when a section was added, the reader should consult the official printed version of the U.S. Code.

SELECTED REFERENCES

[1] Bergeson, Lynn L. (2000). *FIFRA (Federal Insecticide, Fungicide, and Rodenticide Act)*. Basic Practice Series. American Bar Association. 150 p.

[2] CRS Report 96-759, *Pesticide Legislation: Food Quality Protection Act*

of 1996 (P.L. 104-1 70), by Linda-Jo Schierow.

[3] CRS Report RL32218, *Pesticide Registration and Tolerance Fees: An Overview*, by Robert Esworthy.

End Notes

[1] FIFRA also is known as the Act of June 25, 1947.

[2] James L. Beech, U.S. EPA, Office of Pesticide Programs, personal communication, November 20, 2006.

[3] Exceptions are noted in 40 CFR 152.20, 152.25, and 152.30.

[4] Ingredients in pesticide products are categorized as active or inert. Active ingredients are those that are intended to control the pest, while inert ingredients are used to deliver the active ingredients effectively to the pest. Inert ingredients often are solvents or surfactants and often comprise the bulk of the pesticide product. Some inerts are known to be toxic, and some are known to be harmless, but EPA lists most in the category "inerts of unknown toxicity."

[5] That is, use on food crops, animal feed crops, or food products directly (e.g., grains, fruits, or vegetables after harvest).

[6] U.S. House of Representatives, Committee on Commerce, *Food Quality Protection Act of 1996*, H.Rept. 104-669, part 2, 104th Cong., 2nd sess, 1996, p. 6.

[7] Ibid.

[8] EPA. Pesticide Reregistration Facts. October 26, 2006. http://www.epa.gov/oppsrrd1/ reregistration/ reregistration _facts.htm, visited November 20, 2006.

[9] Registrations also may be canceled under other conditions, for example, if data are not submitted in response to EPA's request for additional information to maintain a registration or if a registrant fails to pay the maintenance fee.

In: Pesticide Use in the U.S.: Law and Fee Issues ISBN: 978-1-61122-053-7
Editors: Regina M. Bianchi

Chapter 2

PESTICIDE USE AND WATER QUALITY: ARE THE LAWS COMPLEMENTARY OR IN CONFLICT?

Claudia Copeland

SUMMARY

This report provides background on the emerging conflict over interpretation and implementation of the Federal Insecticide, Fungicide, and Rodenticide Act (FIFRA) and the Clean Water Act (CWA). For the more than 30 years since they were enacted, there had been little apparent conflict between them. But their relationship has recently been challenged in several arenas, including the federal courts and regulatory proceedings of the Environmental Protection Agency (EPA). In this report, a brief discussion of the two laws is followed by a review of the major litigation of interest. EPA's efforts to clarify its policy in this area are discussed, including a regulation issued in 2006 that was subsequently vacated by a federal court, as well as possible options for EPA and Congress to address the issues further.

FIFRA governs the labeling, distribution, sale, and use of pesticides, including insecticides and herbicides. Its objective is to protect human health and the environment from unreasonable adverse effects of pesticides. It establishes a nationally uniform labeling system requiring the registration of all pesticides sold in the United States, and requiring users to comply with the national label. The CWA creates a comprehensive regulatory scheme to control the discharge of pollutants into the nation's waters; the discharge of pollutants without a permit violates the act.

Several federal court cases testing the relationship between FIFRA and the CWA have drawn attention since 2001. In two cases concerning pesticide applications by agriculture and natural resources managers, the U.S. Ninth Circuit Court of Appeals held that CWA permits are required for at least some discharges of FIFRA-regulated pesticides over, into, or near U.S. waters. It held in a third case that no permit was required for the specific pesticide in question. Most recently, the U.S. Second Circuit Court of Appeals ruled in March that a CWA discharge permit for mosquito control activities is not required before April 2011, when EPA is expected to issue a CWA general permit covering such activities.

Several of the rulings alarmed a range of stakeholders who fear that requiring CWA permits for pesticide application activities would present significant costs, operational difficulties, and delays. Pressed to clarify its long-standing principle that CWA permits are not required for using FIFRA-approved products, EPA in 2006 issued a rule to formalize that principle in regulations. Environmental activists strongly opposed EPA's actions, arguing that FIFRA does not protect water quality from harmful pollutant discharges, as the CWA is intended to do. Other stakeholders, such as pesticide applicators, endorsed the rule, although some would like to see its application broadened to include pesticide drift. The EPA rule was challenged, and in January 2009 a federal court vacated the regulation. Several industry groups petitioned for a rehearing by the full Sixth Circuit Court of Appeals, while the federal government asked the court to stay for two years, until April 2011, the order vacating the exemption, to provide time for working with states to develop a general permit for pesticide applications covered by the decision. The court denied the request for rehearing and granted the government's request for a two-year delay. EPA proposed a draft general permit in June. The Supreme Court declined to review the case.

Some believe that the controversy will only be resolved by congressional action to clarify the intersecting scope of the Clean Water Act and FIFRA. Legislation intended to do so by codifying EPA's policy in law was introduced in the 109th Congress, but it was not enacted. For now, it is unclear whether these issues will receive new attention in the 111th Congress.

INTRODUCTION

It has been noted that "[t]he potential for overlapping and potentially conflicting regulatory scope between federal statutes is common, especially in the heavily regulated area of environmental protection."[1] This potential has received attention in connection with implementation of the Federal Insecticide, Fungicide, and Rodenticide Act (FIFRA)[2] and the Federal Water Pollution Control Act (Clean Water Act, CWA).[3] FIFRA requires the Environmental Protection Agency (EPA) to regulate the sale and use of pesticides in the United States through registration and labeling. The CWA is the principal federal law governing pollution in the nation's surface waters.

Pesticides used to control weeds, insects, and other pests receive public attention because of potential impacts on humans and the environment. Depending on the chemical, possible health effects from overexposure to pesticides include cancer, reproductive or nervous-system disorders, and acute toxicity. Similar effects are possible in the aquatic environment. Recent studies suggest that some pesticides can disrupt endocrine systems and affect reproduction by interfering with natural hormones.[4] However, many pesticides and their breakdown products do not have standards or guidelines, and current standards and guidelines do not yet account for exposure to mixtures and seasonal pulses of high concentrations. Effects of pesticides on aquatic life are a concern, because intensive surveys done by the U.S. Geological Survey found that more than one-half of streams sampled had concentrations of at least one pesticide that exceeded an EPA guideline for the protection of aquatic life. Whereas most toxicity and exposure assessments of pesticides are based on controlled experiments with a single contaminant, sampling by the U.S. Geological Survey found that most contamination of waterbodies occurs as pesticide mixtures.[5]

For the more than 30 years since Congress enacted FIFRA and the Clean Water Act, there had been little apparent direct conflict between them. EPA's operating principle during that time had been that pesticides used according to the requirements of FIFRA do not require regulatory consideration under the CWA. EPA had never required CWA permits for use of FIFRA-approved materials, and EPA rules did not specifically address the issue. However, EPA's interpretation and operating practice regarding the relationship between the two laws have recently been challenged in several arenas. Federal courts have been one of two battlegrounds so far where the potential conflict between the regulatory scope of these two laws has been waged. EPA regulatory proceedings have been the second battleground area. Congressional action could add a third testing of the issues.

At issue is how FIFRA-approved pesticides that are sprayed over or into waters are regulated and, specifically, whether the FIFRA regulatory regime is sufficient alone to ensure protection of water quality or whether such pesticide application requires approval under a CWA permit. The issue arose initially over challenges to some routine practices in the West (weed control in irrigation ditches and spraying for silvicultural pest control on U.S. Forest Service lands). It subsequently drew more attention in connection with efforts by public health officials throughout the country to combat mosquito-borne illnesses such as West Nile virus. The litigation created uncertainty over whether application of pesticides and herbicides to waterbodies requires a water discharge permit. EPA tried to promulgate policy to clarify the relationship of the two laws and to address conflicts resulting from several judicial rulings, ultimately in a regulation issued in November 2006. That rule was challenged by multiple parties, and in January 2009, a federal appellate court vacated the rule. A related issue of interest to many pesticide applicators, but not yet addressed by EPA policy or rule, concerns pesticides that unintentionally impact waterbodies through drift or migration from nearby land, such as a field of crops.

This report provides background on the conflict over interpretation and implementation of FIFRA and the Clean Water Act. A brief discussion of the two laws is followed by a review of the major litigation of interest. EPA's efforts to clarify its policy in this area and the November 2006 rule and the 2009 federal court ruling are discussed, as well as possible options for EPA and Congress to further address the FIFRA-CWA issues. In June, EPA proposed a draft general CWA permit that it intends to finalize by April 2011 in response to the court ruling.

THE LAWS

FIFRA is a regulatory statute governing the licensing, distribution, sale, and use of pesticides, including insecticides, fungicides, rodenticides, and other designated classes of chemicals. Its objective is to protect human health and the environment from unreasonable adverse effects of pesticides. To that end, it establishes a nationally uniform pesticide labeling system requiring the registration of all pesticides and herbicides sold in the United States, and requiring users to comply with conditions of use included on the national label. A FIFRA label encompasses the terms on which a chemical is registered, and its requirements become part of FIFRA's regulatory scheme. In registering the chemical, EPA makes a finding that the chemical "when used in accordance with widespread and commonly recognized practice ... will not generally cause unreasonable adverse effects on the environment" (7 U.S.C. § 136a(c)(5)(D)).

EPA reviews scientific data submitted by pesticide manufacturers on toxicity and behavior in the environment to evaluate risks and exposure associated with the pesticide product's use and takes into account the costs and benefits of various pesticide uses. If a registration is granted, the agency specifies the approved uses and conditions of use, which the registrant must explain on the product label. EPA may classify and register a pesticide product for general use or for restricted use (those judged to be more dangerous to the applicator or to the environment which can only be applied by or under the direct supervision of a person who has been trained and certified). FIFRA preempts state, local, and tribal regulations stricter than or different from EPA rules with respect to labeling requirements, but allows states and localities to adopt more restrictive conditions with regard to sale and use.

Use of a pesticide product in a manner not consistent with its label is prohibited, and the law provides civil and criminal penalties for violations. Under FIFRA, EPA generally enforces the law's requirements. However, the law also gives states with adequate enforcement procedures, laws, and regulations primary authority for enforcing FIFRA provisions related to pesticide use.

The objective of the CWA is to "restore and maintain the chemical, physical, and biological integrity of the Nation's waters." To that end, it creates a comprehensive regulatory scheme to control the discharge of waste and pollutants; the discharge of pollutants into waters of the United States without a permit violates the act. The permit requirement is at the heart of the act's compliance and enforcement strategy. Several aspects of these core requirements in the law are important to evaluating whether the CWA applies to specific activities, including whether there is a discharge from a point source (a discrete conveyance such as a pipe, ditch, container, vessel, or other floating craft), whether the discharge is made into waters of the United States, and whether the material discharged is a pollutant; all of these terms are defined in the act. Especially key in the current context is whether pesticides are pollutants under the act. This issue has been central to much of the judicial and regulatory debate over whether the two laws, CWA and FIFRA, are complementary or in conflict. CWA Section 502(6) (33 USC § 1362(6)) defines pollutant thus:

> The term 'pollutant' means dredged spoil, solid waste, incinerator residue, sewage, garbage, sewage sludge, munitions, chemical wastes, biological materials, radioactive materials, heat, wrecked or discarded equipment, rock, sand, cellar dirt and industrial, municipal, and agricultural waste discharged into water.

Section 402 of the act establishes the National Pollutant Discharge Elimination System (NPDES) permitting requirement, which regulates the lawful discharge of pollutants. The act defines "discharge of a pollutant" to mean "any addition of any pollutant to navigable waters from any point source" (CWA Section 502(12); 33 USC § 1362(12)). Discharges are permitted if they are authorized under a NPDES permit that meets CWA requirements, including protecting the receiving waters. NPDES permits specify limits on what pollutants may be discharged and in what amounts. They also include monitoring and reporting requirements. They are either individual case-by-case permits or general permits applicable to similar categories of activities and similar waste discharges. Under the CWA, qualified states issue NPDES permits to regulated sources and enforce permits, and the law allows states to adopt water quality requirements more stringent than federal rules. As of 2009, 46 states have been delegated authority to administer the permit program; EPA issues discharge permits in the remaining states.

The NPDES permit is the act's principal enforcement tool. EPA may issue a compliance order or bring a civil suit in U.S. district court against persons who violate the terms of a permit, and stiffer penalties are authorized for criminal violations of the act. As a practical matter, the majority of actions taken to enforce the law are undertaken by states, both because states issue the majority of permits to dischargers and because the federal government lacks the resources for day-to-day monitoring and enforcement. In addition, individuals may bring a citizen suit in U.S. district court against persons who violate the terms of a CWA-authorized permit or who discharge without a valid permit. FIFRA does not authorize citizen suits.

Throughout the United States, pesticides often are applied in, onto, or near waterbodies to control weeds and insects. Whether those pesticides are adversely affecting water quality has not been a disputed issue until recently. It has been EPA's long-standing practice and interpretation of the laws that a

CWA permit is not required when pesticide application is done in a manner consistent with FIFRA and its regulations. But that interpretation has been challenged in several lawsuits brought since the late 1 990s that have been decided since 2001.

THE LITIGATION

Five federal court cases testing the relationship between FIFRA and the CWA have drawn the most attention, three in the U.S. Court of Appeals for the Ninth Circuit in the West, concerning pesticide applications by agricultural and natural resource managers, and two in the Second Circuit Court of Appeals in the East, involving the use of pesticides by government and public health authorities for mosquito control. These cases have been brought principally under the citizen suit provisions of the CWA. Two of the Ninth Circuit decisions have held that CWA permits are required for at least some activity involving the point source discharge of FIFRA-regulated pesticides over or into waters of the United States, and the third held that a permit was not required because the specific pesticide was not a chemical waste. The Second Circuit ruled in two cases; most recently, it ruled that trucks and helicopters that discharge pesticides are point sources, but it deferred requiring permits until EPA issues a CWA general permit, which is expected in April 2011.

The Ninth Circuit Cases

The first of the major cases on these issues involved application of herbicides in irrigation ditches. In the case, a major issue was whether the application of pesticides constitutes the discharge of a pollutant. Environmental groups challenged application of an aquatic herbicide called Magnicide H to kill weeds and algae and sought to require that the applicator, a municipal corporation that operates a system of irrigation canals in Oregon, obtain an NPDES permit.

The Ninth Circuit Court of Appeals endorsed the lower court's ruling that the pesticide was a pollutant under the CWA, and that the irrigation canals into which the pesticide was being sprayed are "waters of the United States."[6] But it rejected the lower court's holding that a CWA permit was not required because the pesticide was properly regulated by FIFRA and had an EPA-approved FIFRA label. The appeals court ruled that FIFRA and CWA have different purposes and that, as such, neither could be controlling on the application of the other. The court said that FIFRA creates a comprehensive regulatory scheme for the labeling of pesticides, requiring that all insecticides and herbicides sold in the United States be registered with the EPA. It and the CWA have different, although complementary, purposes, the court said, and using a pesticide with a FIFRA-approved label does not obviate the need to obtain a CWA permit. The FIFRA label is the same nationwide. The CWA permit considers local environmental conditions, which the FIFRA label does not. Thus, a nationwide label on a FIFRA-regulated chemical could not be controlling on whether a CWA permit is required, because it does not account for location-specific requirements. The court reversed the district court's grant of summary judgment in favor of the defendants.[7]

Several of the states within the Ninth Circuit subsequently took actions to respond to this ruling. California and Washington amended their water quality program rules to require NPDES permits for pesticide applicators. Oregon did not mandate permits, but suggested that pesticide applicators obtain state-issued permits to protect against lawsuits. Other states outside of the Ninth Circuit have continued their long-standing practice of not issuing permits to persons who apply pesticides to waters of the United States.

The second major case in the West involved an annual U.S. Forest Service (USFS) aerial spray program over national forest lands in Oregon and Washington. Environmental groups filed a lawsuit challenging the spraying program, saying that the environmental impact statement (EIS) prepared by the USFS was inadequate and that the Forest Service had failed to obtain a CWA permit, which they argued is required for this type of aerial spraying. The appeals court reversed the district court's grant of summary judgment for the Forest Service and instructed the lower court to enter an injunction prohibiting the federal agency from further spraying until it acquires an NPDES permit and completes a revised EIS.[8] The court disagreed with the argument of the Forest Service that the spraying is nonpoint source water pollution, which does not require an NPDES permit. The court held that the insecticides meet the CWA definition of "pollutant" and that the application came from an aircraft equipped with spraying apparatus, thus meeting all of the elements of the CWA's definition of point source pollution.

In September 2003, the EPA General Counsel issued a legal memorandum to officials in states located in the Ninth Circuit responding to the *Forsgren* case. The memorandum said that EPA disagreed with the court's holding in the case and that outside the Ninth Circuit, EPA would continue its long-standing interpretation of FIFRA and the CWA. Within the Ninth Circuit, the memo said, EPA would not acquiesce to the ruling in the case of materials other than pesticides (such as those used for fire control), or in circumstances where pesticides are not applied directly over and into waters of the United States.[9]

The third Ninth Circuit case involved an effort by the Montana Department of Fish, Wildlife and Parks to intentionally apply the pesticide antimycin to a river in order to remove non-native trout species and thus to allow re-introducing a threatened fish species into the river. The director of the department was sued under the citizen suit provision of the CWA by a citizen who sought to require the department to obtain an NPDES permit before applying the pesticide.

The court held in this instance that no NPDES permit was required, because the facts of the case demonstrated that, following application as intended, the antimycin dissipated rapidly, leaving no excess portions or residual chemical that should be characterized as chemical waste, and thus is not a pollutant under the act.[10] Intentionally applied and properly performing pesticides are not pollutants, the court said.

The court distinguished this case from its ruling in *Headwaters*, saying that the factual scenarios differ, because "in that case the 'chemical waste' for which a NPDES permit was required was not a pesticide serving a beneficial purpose and intentionally applied to water, but was a chemical that remained in the water after the Magnicide H performed its intended, beneficial function."[11]

Further, the court stated that its analysis accords with EPA's construction of the CWA's definition of "chemical waste" in the context of intentionally applied pesticides, and that the agency's 2003 Interim Statement and Guidance addressing the issue (discussed below) is entitled to some deference. EPA's interpretation as presented in that Interim Statement is reasonable and not in conflict with the expressed intent of Congress, the court said.

The Second Circuit Cases

Two cases in the Second Circuit involved the use of pesticides for mosquito control. In the first case, several residents of the Town of Amherst, NY, sought to halt aerial application of pesticides without a CWA permit. The district court initially dismissed the case, stating that spray drift is not chemical waste under the CWA and that the pesticide use was best regulated under FIFRA. But the appeals court remanded the case to the district court for further development of the record. [12] Although this ruling may not be cited as precedent, it is notable in that, while EPA had filed an *amicus curiae* brief providing its views on this particular case, the court invited EPA to offer its views broadly on the policy and legal questions. The court stated:

> Until the EPA articulates a clear interpretation of current law—among other things, whether properly used pesticides released into or over waters of the United States can trigger the requirement for NPDES permits ...—the question of whether properly used pesticides can become pollutants that violate the CWA will remain open. Participation by the EPA in this litigation in any way that permits articulation of the EPA's interpretation of the law in this situation would be of great assistance to the courts.[13]

The second pertinent case in the Second Circuit also involved the use of pesticides for control of mosquitoes. Plaintiffs in the case, a citizens group, sought an injunction to halt the aerial and ground spraying, arguing that although the pesticides were properly regulated under FIFRA, the spraying program involved the discharge of a pollutant without a CWA permit, and thus was a violation of that law. A federal district court held that FIFRA-compliant spraying activity did not amount to the discharge of a pollutant into navigable waters from a point source, and thus did not violate the CWA. In March, the Second Circuit Court of Appeals disagreed with the district court's finding that trucks and helicopters were not "point source," which are required to have CWA permits in order to discharge lawfully, but this court held that no permits would be required for the challenged activities until EPA issues a general permit in April 2011, as it plans to do (see discussion below, "Options for EPA and Congress"). [14]

Other Litigation

Other lawsuits have followed these cases. For example, private citizens who operate an organic fruit farm in Gem County, Idaho, brought suit against the local mosquito abatement district there, seeking to require a CWA permit for pesticide spraying. Finding itself in the proverbial spot "between a rock and a hard place," the mosquito abatement district applied for a permit from EPA, which the agency declined to issue, based on its long-standing policy and legal interpretation. Thereafter, the mosquito abatement district filed a lawsuit against EPA in an attempt to obtain a declaration that a CWA permit is not needed and to avoid the citizen suit litigation, which is pending in federal court in Idaho. The mosquito abatement district asked the federal court either for a judgment saying that no permit is required or, if the court were to determine otherwise, an order directing EPA to process its CWA permit application. In January 2005, the federal district court in the District of Columbia dismissed the case because the mosquito abatement district and EPA were in agreement that no CWA permit is required for pesticide applications that are consistent with FIFRA.[15]

In other locations, citizen groups have given notice, as required by the CWA, of possible lawsuits to expand the precedent from the Ninth Circuit cases to other types of operations. For example, two actions were threatened in 2004 and 2005 against Maine blueberry farmers for failing to obtain a CWA permit for spraying pesticides that may drift off-target from land into waterbodies. In response to the litigation pressure, however, both farmers subsequently announced plans to cease aerial spraying and instead rely on ground spraying, until such time as government or the courts clarify the law.

EPA'S REGULATORY RESPONSES

The rulings by the Ninth Circuit in the *Talent* and *Forsgren* cases and possible endorsement by other courts greatly alarmed a range of stakeholders in the regulated community, including forestry, agriculture, and pesticide applicators, as well as municipal and public health officials concerned with the need to control mosquitoes and other vectors associated with diseases such as West Nile virus and malaria. They feared that CWA permit requirements would be extended to agricultural and other activities that have not traditionally been regulated under the CWA. They argue that if permits tailored to particular circumstances are deemed necessary, such requirements would present significant costs, operational difficulties, and delays to applicators. They also would put pressure on limited federal and state CWA permitting resources. In their view, requiring permits will not be environmentally helpful, but the expense and long delays of permitting proceedings will hamper programs that are needed for controlling pests that threaten public health and crops. In response, EPA issued two interpretive guidance documents (in 2003 and 2005) and in 2005 proposed a rulemaking to formalize its long-standing position on CWAFIFRA issues. A final rule was promulgated in November 2006 but was vacated by a federal court in 2009, as discussed below.

EPA Guidance

After the *Altman v. Town of Amherst* ruling in 2002, industry, states, and others, including some in Congress, pressed EPA to clarify the emerging conflicts over the two laws. EPA responded in July 2003 with an interim guidance memorandum and a final guidance document in 2005.[16] The agency's consistent position, expressed in both guidance documents, was that pesticides applied in a manner consistent with FIFRA do not constitute either chemical wastes or biological materials under the definition of pollutant in Section 502(6) of the CWA. The rationale for this position was that it is consistent with over 30 years of CWA administration. At the same time, EPA said that pesticide applications in violation of FIFRA, that is, when not used or applied according to applicable labeling requirements, would be subject to all relevant statutes, including the Clean Water Act.

Environmental activists strongly objected to EPA's position in the guidance, which they viewed as contrary to the judicial rulings. These groups reiterated points made by the Ninth Circuit court in the *Headwaters* and *Forsgren* rulings, namely that chemical and biological pesticides are pollutants within the meaning of the CWA, as the law defines pollutants broadly and includes, among other substances, chemical wastes, biological materials, and agricultural wastes. As that court has declared, environmentalists said, FIFRA does not override the CWA, and the two statutes must work in tandem to prevent injury to aquatic life. They also argued that EPA was wrongly deciding that materials with beneficial uses should not be construed as pollutants under the CWA.

Environmentalists' objections also went to the policy problems of relying on FIFRA to protect water quality from pesticide applications, as that would be the result of EPA's position. That position, critics said, turns on whether the pesticide application conforms procedurally with FIFRA requirements, not what is the water quality impact of that pesticide. Other concerns raised by critics included the fact that while the FIFRA registration process calls for ecological risk assessment that may be adequate for producing nationally applicable labels, it does not ensure that local water quality standards are maintained and does not account for additive or synergistic effects of multiple pollutants discharged to a particular waterbody. Environmentalists argued that the CWA provides the means to determine whether, and under what conditions, it is safe to discharge a particular pesticide into a particular body of water, and that FIFRA's nationally uniform labeling system cannot do that. FIFRA is not specifically charged with ensuring the chemical, physical, and biological integrity of U.S. waterways, and satisfaction of a pesticide's FIFRA labeling criteria does not automatically satisfy water quality concerns, as the NPDES permit process is intended to do. They also maintained that FIFRA fails to consider the lasting effects that pesticide residues have on a local ecosystem and that localized analysis of the environmental impact of pollutant discharges under the CWA is necessary, due to the toxic residues that remain after pesticide application, which FIFRA does not address.

Industry welcomed the thrust of the EPA guidance but also urged that it be broadened. Agricultural groups requested that EPA include other classes of applications under the guidance, such as aquaculture and crop production. Beyond the types of uses described in the guidance, some argued that EPA should additionally clarify that CWA permits are not required in the case of pesticides that are applied over land and then inadvertently impact waterbodies through drift and migration. Many requested that EPA address the issues definitively in a rulemaking, rather than in non-binding guidance. In their view, without clear regulatory language supporting EPA's interpretation, pesticide applicators would still face the prospect of citizen lawsuits and NPDES permit requirements.

Many states and local governments, including agriculture agencies, irrigation districts, and mosquito abatement districts, strongly endorsed EPA's proposed clarification of its interpretation of the two laws. However, a few—especially states located in the jurisdiction of the federal Ninth Circuit—expressed a different view. The Oregon Department of Environmental Quality and California State Water Resources Control Board commented that the Interim Statement conflicted with legal precedent in the *Headwaters* case. They urged EPA, if it wishes to create an exemption for pesticide applications conducted in compliance with FIFRA, to ask Congress to amend the Clean Water Act and FIFRA accordingly.

Regulatory Proposal

At the same time that it issued the 2005 guidance, EPA proposed a rulemaking to codify the substance of the guidance in CWA regulations, which it promulgated in November 2006.[17] The final rule, which became effective January 26, 2007, added two specific circumstances that are excluded from NPDES permit requirements, when the application complies with relevant requirements of FIFRA:

- the application of pesticides directly to waters of the United States in order to control pests (e.g., to control mosquito larvae or aquatic weeds); and
- the application of pesticides to control pests that are present over waters of the United States, including near such waters, where a portion of the pesticides will unavoidably be deposited to waters of the United States in order to target the pests effectively.

In the final rule, EPA provided a lengthy discussion of its rationale that pesticides, when applied pursuant to FIFRA, are not chemical wastes or biological materials and thus are not what the CWA defines as "pollutants" (see discussion, page 3). However, EPA also acknowledged that application of pesticides may leave residual materials in U.S. waters after the product has served its beneficial purpose and that these residual materials may be "pollutants" under the act at that later time. Nonetheless, even in such cases, EPA said, the initial application of the pesticide does not require an NPDES permit because EPA does not consider it to be a pollutant *at the time of its discharge* into water.[18] The agency also responded to some public comments on the proposal that had criticized the adequacy of the FIFRA registration process for consideration of water quality, local conditions, etc. EPA said that the "regulatory and non-regulatory tools under FIFRA provide means of addressing water quality problems arising from the use of pesticides," particularly the registration and re-registration processes, which consider both human health and aquatic resource impacts.[19]

Judicial Challenge to the Rule

The 2006 rule prompted multiple lawsuits by industry and environmental groups in almost every judicial circuit nationwide. The litigation was consolidated in the Sixth Circuit Court of Appeals. Industry's challenge argued that the rule is arbitrary and capricious because it treats pesticides applied in violations of FIFRA as pollutants, while treating the same pesticides used in compliance with FIFRA as non-pollutants. It also sought to expand the rule to apply to all pesticides and all agricultural applications of pesticides, including applications to land that drift over or into water. Environmentalists' challenge claimed that, by exempting FIFRA-compliant applications of pesticides from CWA requirements, EPA ignores its duties under the Clean Water Act.

The court's ruling was issued January 7, 2009.[20] EPA had argued that at the time of discharge, a pesticide is a non-pollutant. Excess pesticide or pesticide residues do not exist until after the discharge is complete, EPA said, and therefore should be treated as nonpoint source pollutants that do not require CWA permits. The court rejected EPA's attempt to "inject[] a temporal requirement to the 'discharge of a pollutant,'" and it said that such an interpretation is unsupported by the CWA, and is also contrary to the purpose of the permitting program. The court said, "If the EPA's interpretation were allowed to stand, discharges that are innocuous at the time they are made but extremely harmful at a later point would not be subject to the permitting program." It concluded that "there is no room for the EPA's argument that residual and excess pesticides do not require an NPDES permit," and the court thus vacated the rule. The vacatur was scheduled to take effect April 9, 2009, but subsequently the Sixth Circuit granted the government's request to delay the effective date of the ruling, so that EPA could develop a regulatory response, as discussed below. In the meantime, the rule remains in effect.

In February 2010, the Supreme Court declined to review the Sixth Circuit's ruling.

Congressional Interest and Future Options

Congressional interest in these issues became apparent after the first federal appeals court ruling in the 2001 *Headwaters v. Talent* ruling. Two congressional hearings focused on implications of the cases for pesticide use generally and for local governments' efforts to control mosquito-borne illnesses such as West Nile Virus. Also, a hearing also was held on legislation introduced in the 109th Congress to clarify the scope of the CWA regarding the use of FIFRA-approved pesticides, fire retardants, and biological control organisms.

In October 2002, a House Transportation and Infrastructure subcommittee held a fact-finding hearing on the issues.[21] The subcommittee's particular concern derived in part from the fact that one of the key practices used to manage stormwater runoff, which is regulated under the Clean Water Act, is to collect and hold it in retention ponds, basins, drainage ditches, etc. Such practices can be at odds with the public health objective of controlling insect-breeding habitat by eliminating or draining sources of standing water. Stormwater management practices typically allow collected water to drain slowly, while public health efforts would prefer that it be removed quickly. Another way to address the public health concerns is to spray pesticides on stormwater management structures and other areas of standing waters. The question for this subcommittee was the uncertainty raised by the litigation over the CWA-FIFRA issues for communities, industries, and others needing to maintain stormwater control systems. An EPA official, while acknowledging that the issue of CWA jurisdiction over pesticide spraying is "new territory" for the agency, said that EPA believes there is no inherent conflict between protecting water quality and preventing mosquito-borne disease. At the hearing, some Members and public witnesses urged EPA to provide guidance to resolve uncertainties raised by the court rulings.

A hearing held by a House Government Reform subcommittee in October 2004 examined challenges to controlling West Nile Virus.[22] The hearing was an opportunity for some Members and witnesses to express the view that EPA's July 2003 interim guidance, while helpful in clarifying EPA's position, failed to resolve all legal uncertainty, since it would not bind nonfederal entities or bar citizen lawsuits. Witnesses said that EPA's guidance is a nonbinding legal document that would not deter filing of citizen lawsuits seeking to impose a permit requirement. Supporters of this view urged EPA to settle the legal questions through a formal rulemaking to revise CWA rules. An EPA official said that even if EPA were to promulgate a rule (as it subsequently did), states will still have the discretion to continue to require non-NPDES permits, and a formal rule would not preclude citizen lawsuits from seeking to force localities to file for permits. EPA acknowledged these same points in the 2005 guidance. Others at this hearing agreed on the need for a formal rulemaking, but recommended that in doing so, EPA should reverse the interpretation detailed in the guidance, not codify it.

In the 108th Congress, Senate appropriators included language in their report on EPA's FY2005 funding bill calling on EPA to finalize the interim guidance by December 2004 and to clarify the long-standing distinction between agriculture and silviculture activities that do and do not require CWA permits.[23]

In 2003, a number of House and Senate Members urged the Bush Administration to support Supreme Court review of the *Forsgren* case, but ultimately the Administration did not endorse industry's request for a review, and the Court did not grant certiorari. Some Members of Congress also submitted comments in support of the July 2003 interim guidance document and the January 2005 regulatory proposal.

Options for EPA and Congress

As described above, in January 2009, a federal court rejected EPA's rationale for its 2006 rule that attempted to specify circumstances in which pesticides applied to waters of the United States do not require NPDES permits. The court appeared to leave little room for EPA to fashion a new rule consistent with the agency's long-standing view that FIFRA-compliant applications do not require CWA permits. Agriculture industry groups are fearful that the court's ruling will lead to permit requirements for each pesticide application, placing significant burdens on industry and EPA. Accordingly, several industry groups (the American Farm Bureau Federation, American Forest and Paper Association, and CropLife America, the trade organization for agriculture and pest management) petitioned for a rehearing of the case by the full Sixth Circuit Court of Appeals. The court rejected the rehearing request in August.

The federal government did not seek a rehearing of the case. Instead, the government petitioned the court for a two-year stay of the order vacating the exemption, to give EPA time to work with states and the regulated community to develop a general permit for pesticide applications covered by the decision. State water pollution agencies supported the government's request for the two-year delay, which the court granted on June 8, 2009. As a result, the court's ruling is scheduled to take effect on April 9, 2011.

The two basic types of NPDES permits are individual permits that are specifically tailored for an individual discharger, and general permits that cover categories of point sources having common elements and that discharge the same types of wastes. General permits allow the permitting authority to allocate resources efficiently, especially when there is potentially a large number of permittees, and to provide timely permit coverage. Both individual and general permits are enforceable by the permitting authority and by private citizens (in federal court).

EPA uses its authority to issue NPDES general permits frequently, most recently having issued a general permit to cover discharges incidental to the normal operation of vessels (Vessel General Permit, or VGP) that applies to approximately 69,000 vessels.[24] Typically, dischargers seeking coverage under a general permit are required to submit a notice of intent to be covered by the permit, but this can be modified. For example, in the VGP, EPA provided automatic coverage for about 20,000 of the covered vessels. Still, even with general permits, development and implementation issues will arise, including how EPA will specify applicable discharge limits based on technology available to treat pollutant constituents found in the discharge (i.e., effluent limits), and limits that are protective of the designated uses of the impacted water (i.e., water quality-based effluent limits), as required by the CWA.

In June EPA proposed a draft pesticide general permit that it expects to finalize before the federal court's vacatur takes effect in April 2011.[25] EPA estimated that the universe of activities affected by the court's ruling is approximately 5.6 million applications annually, which are performed by 365,000 applicators, including mosquito and other flying insect pest control; aquatic weed and algae control, aquatic nuisance animal control, and forest canopy pest control. The permit will cover about 500 different pesticide active ingredients that are contained in approximately 3,700 product labels.

The draft permit requires all operators to minimize pesticide discharges to waters by practices such as using the lowest effect amount of pesticide product that is optimal for controlling the target pest. It also requires operators to prepare pesticide discharge management plans to document their pest management practices. Permittees must monitor for observable adverse effects in the treatment area and where the pesticides are discharged to U.S. waters. The permit will not cover agricultural stormwater runoff or irrigation return flow, as these discharges are statutorily exempt from CWA permitting, and it also will not cover terrestrial application to control pests on agricultural crops or forest floors. The EPA general permit will apply in states and areas where EPA is the NPDES permitting authority, but it is expected to be a model for other states to develop their own general permits.[26]

One issue that EPA could address, in addition to developing a general NPDES permit, is pesticide drift, which many stakeholders had urged EPA to address in the final rule. The Federal Register Notice accompanying the rule noted that, at the time, EPA was awaiting advice from a workgroup of its Pesticide Program Dialogue Committee, which could recommend further actions. This committee was established in 1995 as a forum to provide feedback to EPA on various pesticide regulatory, policy, and program implementation issues. It is authorized pursuant to the Federal Advisory Committee Act (FACA), which details requirements for the management and oversight of federal advisory committees to ensure impartial and relevant expertise and advice to EPA and other agencies. In March 2006, the committee convened a Spray Drift Workgroup charged with studying the issue of pesticide drift across water and its accompanying impact on water quality and wildlife. The following year, the workgroup finalized a report that focused on issues related to product labeling, applicator training, and practices and equipment to mitigate drift and adverse effects. In November 2009, EPA proposed new pesticide labels to reduce the drifting of spray and dust from pesticide applications. The agency proposed guidance for pesticide labeling that is intended to describe scenarios when additional language would need to be added to a pesticide label to reduce drift from applications such as on golf courses or parks and noncommercial applications such as residential use on lawns and gardens.[27]

Other options for EPA relate to implementation of FIFRA and procedures used to evaluate the risks of pesticides during the registration process. Environmentalists have argued for some time that EPA's risk review procedures are inadequate because they fail to account for synergistic and additive effects, as well as sub-lethal and indirect effects of pollutants on the environment. In 2003, EPA convened a task force of officials from its pesticide and water quality offices to explore, among other things, whether the agency's pesticide review processes are protective enough to meet water quality standard limits. One outcome of the task force's ongoing review could be changes to implementation of FIFRA in order to address some of these concerns. Thus, in April 2009, EPA officials announced plans to complete a series of white papers on how to harmonize methods used by the agency's Office of Water and the Office of Pesticide Programs for ecological assessment of pesticide chemicals' water quality risks. The white papers are intended to address what officials acknowledge is a gap between the way the CWA and FIFRA approach pesticide risk.

Congress has several options, too, beginning with conducting further oversight of the issues. Another option could involve legislation to revise the CWA and/or FIFRA, and some have urged EPA to ask Congress to legislate a resolution of these possible conflicts.

For example, Congress could enact legislation to clarify that permits are not required for some or all pesticide spraying activity, as favored by members of the pesticide application industry and others, since many of these stakeholders believe that the November 2006 rule did not provide adequate protection from citizen suits. In light of the federal court's vacatur of that rule, these groups may argue that passage of such legislation is even more urgent. Legislation intended to do so was introduced in the 109th Congress (H.R. 1749 and S. 1269, the Pest Management and Fire Suppression Flexibility Act), but it was not enacted. These bills would have provided that NPDES permits are not required for the use of FIFRA-approved pesticides; chemicals, fire retardants, or water used for fire suppression; biological organisms used for plant pest or weed control; or silviculture activities such as timber harvesting that are not currently regulated as point source activities.

As discussed above, EPA's 2006 rule, now vacated by the court, addressed situations in which pesticides are put directly in waters to control pests (e.g., controlling mosquito larvae or aquatic weeds) or cases of pesticides that are present over water and a portion of the pesticide is deposited in the water (e.g., aerial application to a forest canopy where waters of the United States may be present below the canopy). The proposed legislation in the 109th Congress, in addition to codifying these policies, also would have addressed other, broader circumstances that EPA had declined to include in the rule: applications over land areas that may drift over and into waters of the United States, broad exemption of activities for preventing or controlling plant pests or noxious weeds, and use of fire retardants.[28]

In September 2005, a House Transportation and Infrastructure subcommittee held a hearing on H.R. 1749.[29] Witnesses representing a number of sectors that are pesticide users (state foresters, western irrigation districts, and farmers) testified in support of the legislation, saying that it would resolve existing legal uncertainties about permitting. An EPA witness said that the agency's then- proposed rulemaking was intended to reduce uncertainty about the relationship between FIFRA and the CWA. The EPA official did not expressly endorse the legislation, but he said that EPA appreciated congressional efforts to reduce potential confusion over these issues. There was no further action on either H.R. 1749 or S. 1269 during the 109th Congress, and no similar legislation was introduced in the 110th Congress.

Prior to the 2009 federal court ruling that vacated EPA's rule, some environmental activists favored legislation to clarify that NPDES permits *are* required, since they contended that the rule was unlawful. However, no such legislation was introduced. Others argued during this time that legislation is not needed because, in their view, the CWA is clear enough that permits are required for discharge of pesticides from point sources. The federal court's review of the EPA rule supports that view. At the same time, many who would favor legislation to support a narrow view of the CWA's jurisdiction on this issue acknowledge that any such legislative effort would be controversial and could be seen as representing not clarification but, rather, an environmental rollback. The future direction of congressional attention to these issues, in light of the federal court's ruling and EPA's response, is unclear.

End Notes

[1] Randall S. Abate and Matthew T. Stanger, "Pesticides and Water Don't Mix: Addressing the Need to Close a Regulatory Gap Between FIFRA and the CWA," *Environmental Law Reporter News & Analysis*, January 2005, p. 10056.

[2] 7 U.S.C. §§136-136y.

[3] 33 U.S.C. §§1251-1387.

[4] For information, see CRS Report R40 177, *Environmental Exposure to Endocrine Disruptors: What Are the Human Health Risks?*, by Linda-Jo Schierow and Eugene H. Buck.

[5] U.S. Department of the Interior, U.S. Geological Survey, *The Quality of Our Nation's Waters, Nutrients and Pesticides*, USGS Circular 1225, 1999, pp. 3-9.

[6] In view of the Supreme Court's June 2006 decision in *Rapanos v. United States* (547 U.S. 715, 2006), coverage of irrigation canals as "waters of the United States" may depend on case-specific circumstances, because the Court's plurality opinion in this case made specific reference to *Headwaters, Inc. v. Talent Irrigation District*. For additional information, see CRS Report RL3 3263, *The Wetlands Coverage of the Clean Water Act (CWA) Is Revisited by the Supreme Court: Rapanos v. United States*, by Robert Meltz and Claudia Copeland.

[7] Headwaters, Inc. v. Talent Irrigation District, 243 F.3d 526 (9th Cir. 2001).

[8] League of Wilderness Defenders/Blue Mountains Biodiversity Project v. Forsgren, 309 F.3d 1181 (9th Cir. 2002).

[9] Robert Fabricant, EPA General Counsel, "Interpretive Statement and Guidance Addressing Effect of Ninth Circuit Decision in *League of Wilderness Defenders v. Forsgren* on Application of Pesticides and Fire Retardants," memorandum, September 3, 2003, 7 pp.

[10] Fairhurst v. Hagener, 422 F.3d 1146 (9th Cir. 2005).

[11] Ibid. at 1150.

[12] Altman v. Town of Amherst, N.Y., 47 Fed. Appx. 62 (2d Cir. 2002).

[13] 47 Fed. Appx. at 67.

[14] Peconic Baykeeper Inc. v. Suffolk County, 2d Cir., No. 09-97-cv, March 30, 2010.

[15] Gem County Mosquito Abatement District v. EPA, 398 F. Supp. 2d 1 (D.D.C. 2005)

[16] U.S. Environmental Protection Agency, "Application of Pesticides to Waters of the United States in Compliance With FIFRA, proposed rulemaking and notice of interpretive statement," 70 *Federal Register* 5093, February 1, 2005

[17] U.S. Environmental Protection Agency, "Application of Pesticides to Waters of the United States in Compliance With FIFRA," Final Rule, 71 *Federal Register* 68483, November 27, 2006.

[18] Id. at 68487.

[19] Id. at 6848 8-68489.

[20] National Cotton Council of America v. U.S. Environmental Protection Agency, 553 F.3d 927 (6th Cir. 2009).

[21] U.S. Congress, House, Committee on Transportation and Infrastructure, Subcommittee on Water Resources and Environment, "West Nile Virus: The Clean Water Act and Mosquito Control," Hearing, October 10, 2002, 107th Cong., 2nd sess., unpublished.

[22] U.S. Congress, House, Committee on Government Reform, Subcommittee on Energy Policy, Natural Resources and Regulatory Affairs, "Current Challenges in Controlling the West Nile Virus," Hearing, October 6, 2004, 108th Cong., 2nd sess. (Serial No. 108-274), 182 p.

[23] U.S. Senate, Committee on Appropriations, "Departments of Veterans Affairs and Housing and Urban Development, and Independent Agencies Appropriations Bill, 2005," report to accompany S. 2825, 108th Cong., 2nd sess., pp. 110- 111.

[24] U.S. Environmental Protection Agency, "Final National Pollutant Discharge Elimination System (NPDES) General Permit for Discharges Incidental to the Normal Operation of a Vessel," 73 *Federal Register* 79473-7948 1, December 29, 2008.

[25] U.S. Environmental Protection Agency, "Draft National Pollutant Discharge Elimination System (NPDES) Pestice General Permit for Point Source Discharges from the Application of Pesticides," 75 *Federal Register* 31775-31785, June 4, 2010.

[26] The CWA authorizes EPA to delegate NPDES permitting authority to qualified states, and EPA has done so for the majority of states. For this permit, EPA will be the permitting authority in Massachusetts, New Mexico, Oklahoma, Alaska, Idaho, the District of Columbia, and for Indian lands in a number of states.

[27] U.S. Environmental Protection Agency, "Pesticides; Draft Guidance for Pesticide Registrants on Pesticide Drift Labeling," 74 *Federal Register* 57 166-57168, November 4, 2009.

[28] Reflecting a different approach, Rep. Goodlatte introduced a bill in the 107th Congress, H.R. 5329, that proposed amending FIFRA to expand the definition of what constitutes a "public health" pesticide in an effort to ease industry's ability to register pesticides for use in combating mosquito-borne illnesses.

[29] U.S. Congress, House, Committee on Transportation and Infrastructure, Subcommittee on Water Resources and Environment, "Hearing, 'H.R. 1749, Pest Management and Fire Suppression Flexibility Act,'" 109th Cong., 1st sess., September 29, 2005 (109-33), 110 pp.

In: Pesticide Use in the U.S.: Law and Fee Issues ISBN: 978-1-61122-053-7
Editors: Regina M. Bianchi

Chapter 3

PESTICIDE REGISTRATION AND TOLERANCE FEES: AN OVERVIEW

Robert Esworthy

SUMMARY

The U.S. Environmental Protection Agency (EPA) is responsible for regulating the sale, use, and distribution of pesticides under the authority of two statutes. The Federal Insecticide, Fungicide, and Rodenticide Act (FIFRA) (7 U.S.C. 136-136y), a licensing statute, requires EPA to review and register the use of pesticide products. The Federal Food, Drug, and Cosmetic Act (FFDCA) (21 U.S.C. 346a) requires the establishment of maximum limits (tolerances) for pesticide residues on food in interstate commerce. EPA was also required to reevaluate older, registered pesticides (i.e., "reregistration" for pesticides registered prior to 1984, and more recently, registration review) and to reassess existing tolerances (i.e., tolerance reassessment) to ensure they meet current safety standards. Although U.S. Treasury revenues cover most costs for administering these acts, fees paid by pesticide manufacturers and other registrants have supplemented EPA appropriations for many years as a means of increasing the pace of the agency's activities under FIFRA and FFDCA.

The Pesticide Registration Improvement Act (PRIA 1), included in the Consolidated Appropriations Act, 2004 (P.L. 108-199, Title V of Division G), enacted on January 23, 2004, amended FIFRA and modified the framework for collecting fees to enhance and accelerate the agency's pesticide licensing (registration) activities. The amendments included reauthorization of maintenance fees primarily to support activities related to existing registrations, and established registration service fees to be submitted with applications for new registrations. The Pesticide Registration Improvement Renewal Act, or PRIA 2 (P.L. 110-94), enacted October 9, 2007, reauthorized and revised these fee provisions, which would have expired at the end of FY2008. In February 2010, EPA reported the completion of 1,570 registration and reregistration decisions subject to PRIA during FY2009, for a total of 7,530 since the enactment of PRIA 1 in 2004. For FY2009, EPA reported expending $18.5 million of the $25.5 million received in the form of new registration fees collected in FY2009 ($16.1 million in net receipts) and carried forward from FY2008 ($9.4 million).

Authority for collecting pesticide fees dates back to the 1954 FFDCA amendments (P.L. 518; July 22, 1954), which, as passed, required the collection of fees "sufficient to provide adequate service" for establishing maximum residue levels (tolerances) for pesticides on food. Authority to collect fees was expanded with the 1988 FIFRA amendments (P.L. 100-532). The 1996 amendments to FIFRA and FFDCA, or the Food Quality Protection Act (FQPA) (P.L. 104-170), extended EPA's authority to collect certain fees through FY2001. Congress extended this authority annually through appropriations legislation prior to the enactment of PRIA in 2004.

The FY1998-FY2004 President's budget requests included proposals to modify existing fee structures to further increase revenues for pesticide activities. These proposals were not adopted in legislation and in some cases specifically prohibited by Congress. In each fiscal year budget request since PRIA was enacted in 2004, EPA has included proposals to further increase pesticide fees beyond those authorized. These proposals were not adopted by Congress in each year through FY2010. The FY2011 President's budget request included similar proposals to provide for the collection of additional pesticide fees.

INTRODUCTION

The collection of fees to support U.S. Environmental Protection Agency (EPA) pesticide program activities has been a complex issue for more than 20 years. Authorities to collect fees in addition to appropriated funds have been provided over the years in part to accelerate the agency's review efforts and to fund its increasing statutory responsibilities. Current and past Administration proposals to modify and significantly increase pesticide fees have been at odds with the views of a range of stakeholders and controversial in Congress. Congress acted to address the issues of concern through pesticide fee provisions included in the Consolidated Appropriations Act of FY2004, enacted on January 23, 2004 (P.L. 108-199). This authority for collection of pesticide fees would have expired at the end of FY2008 (with phase-out authority at reduced levels for FY2009 and FY20 10). Enacted October 9, 2007, P.L. 110-94—the Pesticide Registration Improvement Renewal Act (referred to as PRIA 2)[1]—revised and reauthorized the pesticide fee collection provisions effective retroactively to the beginning of FY2008 through FY20 12. PRIA 2 passed in both houses of Congress by unanimous consent.

General U.S. Treasury revenues are used to cover most of the administrative costs of EPA's pesticide program, which implements requirements under the Federal Insecticide, Fungicide, and Rodenticide Act (FIFRA) (7 U.S.C. 136-136y) and the Federal Food, Drug, and Cosmetic Act (FFDCA) (21 U.S.C. 346a), as amended. However, fees also have been imposed on those who manufacture and distribute pesticides in U.S. commerce (i.e., registrants[2]) to supplement EPA appropriations. Provisions in the Consolidated Appropriations Act of FY2004, which became known as the Pesticide Registration Improvement Act of 2003 (PRIA or PRIA 1), modified existing pesticide fee authority to support specified activities and process improvements in an effort to achieve more timely completion of EPA's statutory obligations under the authority of FIFRA and FFDCA. PRIA 2 (P.L. 110-94) renews this authority with some technical revisions, primarily modifications to the fee payment process and an expansion of the range of categories of pesticide registration (licensing) activities subject to fees.

Congress has granted EPA authority to collect fee revenues as a means of accelerating the pace of the agency's activities to meet its statutory obligations required under FIFRA and FFDCA. These activities include review of the science when evaluating new pesticide registrations, and the establishment of the maximum residue allowance (a "tolerance") as necessary. Also included is EPA's reevaluation of older pesticide registration and existing tolerances (i.e., tolerance reassessment) to ensure they meet current standards for protecting human health and the environment (see more detailed discussion in the following section of this report).

In February 2010, EPA released its annual PRIA progress report.[3] *Implementing the Pesticide Registration Improvement Act (PRIA) - Fiscal Year 2009* (or EPA's FY2009 PRIA implementation report) provides information about the registration process, including the status of its registration and reregistration[4] activities, as well as EPA's efforts to improve the processes. EPA reported the completion of 1,570 decisions subject to PRIA in FY2009, compared with 1,677 at the end of FY2008, 1,620 during FY2007, 1,347 during FY2006, 1,098 during FY2005, and 208 decisions completed during FY2004.

The following sections of this report provide a historical overview of federal authority regarding pesticide fees, including the amount of fee revenues collected over time, and summarizes the key elements of PRIA and the revisions reflected in PRIA 2. For a more complete overview of the federal pesticide laws, refer to CRS Report RL3 1921, *Pesticide Law: A Summary of the Statutes*, by Linda-Jo Schierow.

BACKGROUND

FIFRA is a licensing statute that requires EPA to register pesticide products before they can be sold, used, and distributed within the United States. EPA evaluates proposed pesticide registrations under a set of science-based safety standards. Before a registration can be granted for a "food use" pesticide, FFDCA[5] requires that a tolerance (the maximum amount of pesticide residue permitted in or on food and feed) or tolerance exemption be in place.

Under the standards introduced by the 1996 amendments to FIFRA and FFDCA (the Food Quality Protection Act or FQPA; P.L. 104-170), EPA establishes tolerances through rulemaking based on risk assessments and human health criteria to ensure a "reasonable certainty of no harm." For pesticides that are not used on food, FIFRA requires EPA to determine whether and under what conditions the proposed pesticide use would present an unreasonable risk to human health or the environment. EPA was also required to reevaluate older, registered pesticides (i.e., reregistration) and to reassess existing tolerances (i.e., tolerance reassessment)[6] to ensure they meet current safety standards. Congress has amended FFDCA and FIFRA over time to authorize the collection of fees to supplement appropriated funds for these pesticide review activities.

The 1954 amendments to FFDCA[7] authorized the collection of fees to provide adequate service for establishing maximum allowable residue levels (tolerances) for pesticides on food, and they remain the basis for current "tolerance fee" authority. Congress amended FIFRA in 1988 (P.L. 100-532), authorizing the collection of a one-time "reregistration fee" and, through FY1997, annual "maintenance fees" in an effort to accelerate reregistration (review of pesticides registered before 1984).

In the 1996 amendments to FIFRA and FFDCA (FQPA; P.L. 104-170), Congress, concerned with the continued pace of reregistration, extended EPA's authority to collect the annual maintenance fees through FY200 1. In addition, in an attempt to provide resources to address increased responsibilities of implementing new safety standards introduced with the 1996 amendments, maintenance fee authority was expanded to allow a portion of the collected revenue to be used to support the reevaluation of "old" existing tolerances (tolerance reassessment). These pesticide maintenance fees, along with tolerance fees based solely on petitions for establishing new tolerances, were the only pesticide fees collected by EPA during the eight years (FY1996- FY2003) prior to the enactment of PRIA. (A more detailed overview of fee authorities and revenues collected is presented in "A Historical Overview of Pesticide Fee Authorities," later in this report.)

In the annual fiscal year budget requests, the previous two Administrations proposed modifications to the fee structure to significantly increase revenues, primarily to obtain supplemental resources to support increased administrative costs associated with implementing the requirements of FQPA. Proposals generally focused on finalizing a 1999 EPA proposed rule[8] to substantially revise tolerance fees and on a recommendation that Congress discontinue the legislative prohibition on pesticide registration fee authority[9] promulgated in 1988. The current Administration's requests have included similar proposals to allow for the collection of pesticide tolerance fees, as well as proposals for a new fee schedule for registration service fees to increase the overall amount of fee revenue collected, and proposals to raise the current limit for the amount of maintenance fees that can be collected annually.

Shortly after its promulgation, the final 1988 pesticide registration fee regulation was challenged in court by the Chemical Specialties Manufacturers' Association,[10] which questioned the appropriateness of the statutory authority cited. Collection of these registration fees, as promulgated, was temporarily suspended through FY1997 by the 1988 amendments to FIFRA (Section 4[i] [6]). Collecting registration fees as promulgated in 1988 continued to be prohibited subsequently by the 1996 FIFRA/FFDCA amendments (FQPA) and in provisions of annual appropriations bills, including the PRIA provisions in the FY2004 Consolidated Appropriations.

A proposed 1999 regulation to restructure the collection of tolerance fees met with similar resistance. Industry groups questioned the authority to expand fee collection under FFDCA[11] and the lack of a clearly defined schedule of specific agency activities to be supported by fee revenues. These groups also generally opposed the EPA's justification for proposing a tenfold increase, requiring retroactive fee payments, and imposing fees for inert ingredients.[12] Congress initially prohibited promulgation of the tolerance fee rule in EPA's FY2000 appropriations (P.L. 106-377). Similar proposals to increase tolerance fees in EPA's annual budget requests from FY200 1 to FY2004 were prohibited through appropriations legislation.

PRIA as enacted in 2004 (PRIA 1) specifically prohibited collection of *any* tolerance fees, and temporarily replaced (essentially prohibited) other fee authority through FY2008. Despite this prohibition, similar proposals to authorize collection of additional tolerance fees and other pesticide fee revenue increases were included in the FY2005, FY2006, FY2007, FY2008, FY2009, and FY20 10 budget requests. These proposals were not adopted by Congress for those fiscal years. Language contained in the FY2005 supplemental appropriations for military funding enacted May 11, 2005 (P.L. 109-13 § 6033), specifically banned EPA from going forward with rulemaking for collecting pesticide tolerance fees. PRIA 2 (P.L. 110-94), enacted October 9, 2007, continued the prohibition of other fee authority through FY20 12. The FY20 11 President's budget request includes similar proposals to expand the collection of pesticide fees. [13]

The following section summarizes the key provisions of PRIA 2 as enacted.

KEY PROVISIONS OF THE PESTICIDE REGISTRATION IMPROVEMENT RENEWAL ACT AS ENACTED IN 2007

The Pesticide Registration Improvement Renewal Act (P.L. 110-94) or PRIA 2—effective retroactively to October 1, 2007[14]—amended and reauthorized provisions of the Pesticide Registration Improvement Act of 2003 (P.L. 108-199),[15] or PRIA 1. PRIA 1 had amended FIFRA and modified the framework for collecting fees to enhance and accelerate EPA review of pesticide registration and reregistration applications, temporarily superseding the 1988 registration fee authority[16] and suspending tolerance fee authority under FFDCA through FY2008.

As enacted in 2004, PRIA 1 seemed to address many of the issues associated with previously proposed modifications of pesticide fees, and received the support of a large cross section of stakeholders, including organizations representing manufacturers and formulators, agricultural producers, and environmental and public interests.[17] These groups jointly favored the acceleration of EPA's decision process, the simplification of the fee authority, and the detailed schedule of activities determining the allocation of fees collected. The changes reflected in PRIA 2 have generally continued to receive similar support from various groups.[18]

PRIA 2 amended certain PRIA 1 provisions under FIFRA, most notably the addition of 40 new registration application categories and clarification to existing categories, and changes to small business fee waiver options. PRIA 2 also extended the baseline budget protection for the Office of Pesticide Programs (OPP); if congressional appropriations fall below the amount allocated to OPP in FY2002 by more than 3%, authority to assess fees is suspended. In addition, PRIA 2

- extended authority to collect maintenance fees through FY20 12;
- renewed authority for EPA to collect "registration services fees," which would be phased out at the end of FY2014;
- continued the prohibition on the collection of any tolerance fees through FY2012;
- amended the authority for use of funds in the Reregistration and Expedited Processing Fund to include use for registration review;
- required pesticide registrations to be reviewed every 15 years;[19] and
- required EPA to identify reforms to the pesticide registration process to substantially reduce the decision review period.

Reauthorization of Pesticide Registrant Maintenance Fees

Under PRIA 2, the annual maximum maintenance fees per registrant, and in aggregate, remain constant for each of the fiscal years FY2008 through FY20 12, simplifying the formulae under PRIA 1.[20] The annual statutory aggregate limit is $22.0 million for each of the fiscal years FY2008 through FY20 12.[21] PRIA 2 amended FIFRA, increasing the annual maximum fee for registrants with *less than* 50 pesticide registrations to $71,000 for each of the fiscal years FY2008 through FY20 12; or $50,000 if a registrant is a small business (as redefined in PRIA 2). The annual maximum each fiscal year FY2008 through FY2012 for registrants with *more than* 50 registrations is changed to $123,000; $86,000 if a registrant is a small business. Waivers continue to be available for public health pesticides.

Maintenance fees continue to be assessed on existing pesticide registrations to fund registration reviews and tolerance reassessment. The 1996 FQPA placed greater emphasis on inert ingredients and clarified that these chemicals are covered by the definition of a pesticide chemical under FFDCA (§ 201(q)(1)), but not FIFRA. Therefore, EPA must make a determination regarding the establishment of tolerances for inert ingredients. PRIA 2 extended the authority to collect maintenance fees so as to explicitly designate the use of a portion (between 1/8 and 1/7) of the annual aggregate maintenance fees collected for

- the expedited processing of proposed new products that are "similar" or identical to existing products,[22]
- proposed label amendments that require no review of scientific data,
- proposed registrations of public health pesticide uses, and
- the review and evaluation of new "inert" ingredients.[23]

Registration Services Fees

PRIA 1 established registration "services" fees that apply only to new pesticide applications (submitted on or after the effective date of PRIA), with transitional allowances for pending applications. PRIA 2 extended this authority in a new FIFRA Section 33 (7 U.S.C. § 136w-8). These fees are expected to cover a portion of the cost for review and decision-making associated with a registration application, including associated tolerance determinations. As defined initially under PRIA 1, these costs include EPA staff, contractors, and advisory committees engaged in relevant activities for pesticide applications, associated tolerances, and corresponding risk and benefits information and assessment. Under PRIA 2 authority to collect service fees ends at the end of FY2012, with phase-out authority at reduced levels through FY2014.

The category or type of application, the amount of the pesticide registration service fee, and the corresponding decision review time frame in which the agency is to make a decision are prescribed in the act. The EPA Administrator was directed to publish a detailed schedule of covered pesticide applications and corresponding registration service fees, as reported in the July 31, 2007, *Congressional Record* (S 10409 through S 10411).[24] EPA published a revised pesticide service fee schedule October 30, 2007.[25] The amount of the fees varies depending on the specific "service" required. PRIA 2 includes mandatory adjustments of the registration service fees—a 5% increase beginning with registration applications received October 1, 2008 through September 30, 2010,[26] and an additional 5% increase of the service fee in effect as of September 30, 2010. EPA provides a website, including a "decision tree" for determining fee amounts, to assist applicants in identifying the category of their application and the amount of the required fee prior to submission of their application for a pesticide registration or certain tolerance action.[27] During FY2009, EPA revised the fee category interpretations and developed additional guidance on the type of application that fell into each specific fee category.[28] Following stakeholder review, EPA posted the fee category interpretations on the agency's PRIA 2 website.

PRIA 2 modified the existing authorizations for waivers or reductions of registration service fees for minor uses or small businesses under Section 33(b)(7)F of FIFRA, and for partial fee refunds when applications are withdrawn or at the Administrator's discretion.[29] PRIA 2 amended the small business waiver provisions such that a 100% waiver is no longer authorized; a qualified small business is eligible for a partial waiver of 50% or, in some cases, 75% of the registration service fee available under the new provisions. According to its FY2009 PRIA Implementation Report,[30] EPA granted 306 fee waivers/exemptions and denied 6 of the 320 requests reviewed during FY2009 (review was pending for the 8 remaining requests at the end of the fiscal year). The waivers granted included 189 for 75% small business waivers (6 denied) and 76 for 50% waivers. Comparatively, during FY2008, EPA granted 293 of 303 waiver applications reviewed; 176 for 75% small business waivers, 64 for 50% waivers, and 9 for 100% small business waivers that had been received prior to the PRIA 2 effective date.[31] During FY2007, 258 of the total 374 waivers granted were 100% small business waivers. The total of $6.9 million in fees waived or exempted in FY2009 was a decrease from the $8.2 million in fees waived in FY2008 and $11.4 million in FY2007.[32]

Pesticide Registration Fund

As established under PRIA 1, PRIA 2 retains the Pesticide Registration Fund ("the fund") in the U.S. Treasury, to be made available to EPA for purposes defined in the act, without fiscal year limitation.[33] PRIA 2 amended certain provisions regarding the fund, including a requirement that approximately 1/17 of the amount in the fund, but not less than $1 million of the total amount in the fund, be used to enhance scientific and regulatory activities for worker protection for FY2008 through FY2012. PRIA 1 had required a range of $750,000 to $1 million for worker protection activities. Additionally, a new provision in PRIA 2 requires $750,000 for each of the years FY2008 and 2009, and $500,000 for each of the years FY2010 through 2012 is to be used for "Partnership Grants," for projects supporting pesticide risk reduction. Another $500,000 for each of the years 2008 through 2012 is to be used to carry out a "pesticide safety education program."

To ensure that the appropriated funds are not reduced in lieu of fee revenues, PRIA 2 extended the prohibition on authorizing registration services fees unless the amount of congressional appropriations for specified functions conducted by the EPA Office of Pesticide Programs remains no less than 3% below the corresponding FY2002 appropriation.[34] PRIA 2 also continued to stipulate that the authorization to collect and obligate fees must be provided in advance in appropriations acts.[35] These requirements have been met in EPA appropriations for FY2004 through FY2010.[36]

Prohibition of "Tolerance Fees"

PRIA 2 continues to prohibit EPA from collecting "any" tolerance fees under the authority of section 408(m)(l) of FFDCA.[37] Authority for collecting tolerance fees dates back as far as the 1954 amendments to FFDCA (P.L. 518; July 22, 1954), which, as passed, required the collection of fees "sufficient to provide adequate service" for establishing maximum residue levels (tolerances) for pesticides on food (see more detailed discussion below under "A Historical Overview of Pesticide Fee Authorities"). Under PRIA 2, fee revenues to support tolerance assessments are allocated from maintenance fees (for tolerance reassessments) and registration service fees (for new and amended tolerances), through FY2012.

Prohibition of Other Pesticide Fees

PRIA continues to prohibit "other fees" by amending FIFRA Section 4(i)(6) and by replacing 2010 with 2014.[38] Specifically, the collection of fees under the registration fee authority codified in 1988 (Subpart U of CFR part 152) is temporarily replaced and essentially prohibited by this provision.

Reregistration and Expedited Process Fund

PRIA 2 amended FIFRA and expanded the authorization of the use of moneys collected and deposited in the previously established Reregistration and Expedited Process Fund.[39] The use of this fund was expanded to include offsetting costs of "registration reviews under section 3(g)."[40] Prior to this amendment, money in the fund was only to be used to offset the cost of reregistration[41] and for the expedited review of inert ingredients. PRIA 2 established a deadline of October 1, 2022, for EPA to complete registration review decisions for all pesticides registered as of October 1, 2007.[42]

EPA completed the reregistration of active ingredients that require (food) tolerances or exemptions from tolerances, required to be completed by August 3, 2006, as mandated under FQPA (P.L. 104-170, Title IV, § 405). PRIA 1 had extended the statutory deadline for completing reregistrations for active ingredients that do not require tolerances to October 3, 2008. EPA completed its review at the end of September 2008. EPA's final rule for the registration review program as published in the August 9, 2006, *Federal Register*, replaced the agency's pesticide reregistration and tolerance reassessment programs as they approached completion.[43] The agency began implementing the registration review program at the start of FY2007. (See more detailed discussion later in this report "Registration Review").

Reporting Progress under PRIA

PRIA 2 extended the requirement that EPA publish an annual report describing relevant actions taken during each fiscal year, through March 1, 2014. PRIA 2 also retained, with some technical modifications, specific elements to be included in the report. Examples of these reporting elements include progress made in carrying out its obligations under the act, a description of the staffing and resources related to the costs associated with the review and decision-making pertaining to applications, and the progress in meeting the goals for registration review and reregistration timeline requirements. A key provision of PRIA 1 in conjunction with the increased fee revenues was the requirement for EPA to identify reforms[44] to the agency's pesticide registration process with the intent of reducing the current decision review period. EPA released reports in March 2005, March 2006, March 2007, March 2008, and March 2009, and its most recent report, *Implementing the Pesticide Registration Improvement Act (PRIA) - Fiscal Year 2009*, was published February 26, 2010.[45] These reports summarize improvements and accomplishments for each of the fiscal years. (See discussion under "Pesticide Registration and Reregistration Activities Since the Enactment of PRIA").

A Historical Overview of Pesticide Fee Authorities

Various changes and proposed changes to pesticide fee authority led up to the 2004 enactment of PRIA 1. Fees collected by EPA over time to support the pesticide program have included tolerance fees, registration fees, reregistration fees, and maintenance fees. Between 1996 and 2004, EPA collected tolerance fees, primarily for the establishment of pesticide residue limits (tolerances) on food, and maintenance fees, primarily for reregistration reviews and reassessment of existing tolerances. **Table 1** below provides a timeline of key pesticide fee authorities and implementation regulations; the following sections provide a brief description of these actions.

FIFRA and FFDCA Pesticide Fee Collection Authority

Authority for the collection of pesticide fees dates back as far as the 1954 amendments to FFDCA.[46] At the time, § 408(o)[47] required the collection of fees to cover the costs of establishing maximum residue levels ("tolerances") for pesticides on food. Until 1988, tolerance fees were the only pesticide fees collected by EPA. The 1988 amendments to FIFRA (P.L. 100-532) extensively expanded pesticide fee authority. The amendments included a nine-year schedule to accelerate the process of reregistration. To help defray the costs of the accelerated process, EPA was authorized to collect a one-time reregistration fee from producers for their pesticide active ingredients registered prior to 1984, and annual maintenance fees from pesticide registrants through FY1997, for each registered pesticide product. The amounts of fees per registrant were tiered, depending on the number of registrations per registrant, as determined by EPA each fiscal year.

Table 1 . Timeline of Key Legislation and Regulation Regarding Pesticide Fees

Year	Legislation/Regulation	Pesticide Fee Authority/Action
1952	Independent Appropri-ations Act of 1952 (IOAA; 31 U.S.C. § 9701)	Authorizes the head of each agency to prescribe regulations establishing a charge for a service or thing of value provided by the agency.
1954	Federal Food Drug and Cosmetic Act, amended (FFDCA; P.L. No. 518, 21 U.S.C. § 346 [a])	Authorizes fees to accompany initial or modified petitions for establishing tolerances under FFDCA § 408 (o).
1986	EPA Registration Fee Regulation: Proposed (51 *Federal Register* 42974, Nov. 26, 1986)	Proposed a schedule of fees to accom-pany pesticide registration and experimental use permit applications, citing the authority of IOAA.
1988	EPA Registration Fee: Final Regulation (40 CFR 152[u] and 40 CFR 172)	Establishes fees to accompany pesticide registration and experimental use permit applications; authority suspended by the FIFRA amendments passed later that same year (1988).
	Federal Insecticide, Fungicide and Rodenticide Act, amended (FIFRA; P.L. 100-532)	Authorizes reregistration and expedited processing fund: a one-time "reregist-ration" fee and annual "maintenance" fees through FY1997. Prohibited collection of other fees (including "registration fees" as defined in 40 CFR 152[u] and 40 CFR 172).
1996	Food Quality Protection Act (FQPA) (P.L. 104-170): FIFRA and FFDCA, amended	Extends authorization for maintenance fees through FY2001. FFDCA authority (§ 408[m]) amended to cover costs of all tolerance activities and directs EPA to deposit funds collected as maintena-nce fees to be used for reassessing existing tolerances as needed. Prohibits

Table 1. (Continued)

Year	Legislation/Regulation	Pesticide Fee Authority/Action
		collection of registration fees as defined in 40 CFR 152(u) and 40 CFR 172) through FY2001.
1999	EPA Tolerance Fee Rule: Proposed (64 *Federal Register* 31039-31050, June 9, 1999)	Proposed establishment of a tenfold increase in existing tolerance fees and new "tolerance reassessment" fees, including fee for reviewing tolerances for inert ingredients. Fees, to be collected retroactively from 1996, would supplement authorized maintenance fees.
	FY2000 EPA Appropriations (P.L. 106-377)	Prohibited promulgation of a final tolerance fee rule based on EPA's 1999 proposal.
2000	FY2001 EPA Appropriations (P.L. 106-74)	Continued prohibition on promulgation of a final tolerance fee rule as proposed in 1999.
2001	FY2002 EPA Appropriations (P.L. 107-73)	Continued the prohibition on promulg-ation of a final tolerance fee rule based on the 1999 proposal and on collection of registration fees as codified in 1988. Maintenance fees reauthorized and aggregate limit increased.
2002	Farm Security Act (P.L. 107-171)	Senate-proposed pesticide fee authorities considered and deleted in Conference. Conferees questioned the legal basis for EPA's June 9, 1999, proposed rule (64 FR 31039) to collect tolerance fees retroactively and encouraged EPA to withdraw the proposal. (H.Rept. 107-424).
2002-2003	EPA Appropriations: FY2003 (P.L. 108-10) and FY2004 Continuing Resolution (P.L. 108-135; through Jan. 31, 2004)	Prohibited promulgation of a final tolerance fee rule based on the 1999 proposal. Continued prohibition of the collection of registration fees as codified in 1988. Maintenance fees reauthorized; maximum aggregate levels increased.
	S. 1664 and H.R. 3188, proposed; the basis for PRIA 1 provisions later	Would have authorized new a registra-tion service fee, reauthorized mainten-ance fees, required pesticide regulation

Table 1. (Continued)

Year	Legislation/Regulation	Pesticide Fee Authority/Action
	included in the FY2004Consolidated Appropriations Bill (P.L. 108-199)	process reforms, and prohibited collection of tolerance fees.
2004	FY2004 Consolidated Appropriations Act (P.L. 108-199; Division G, Title V), enacted Jan. 23, 2004	Authorized new registration "service" fee, reauthorizes maintenance fees, requires pesticide regulation process reforms, and prohibits the collection of tolerance fees.
	FY2005 Consolidated Appropriations Act (P.L. 108-447), enacted Dec. 8, 2004	Provided continued authorization for the collection of pesticide fees during FY2005 pursuant to P.L. 108-199.
2005	FY2005 supplemental appropriations for military funding (P.L. 109-13, § 6033) enacted May 11, 2005	Banned EPA from going forward with rulemaking for collecting pesticide tolerance fees as prohibited by PRIA.
2006	Interior, Environment, and Related Agencies Appropriations[a] Act for FY2006 (P.L. 109-54), enacted August 25, 2005	Provided continued authorization for the collection of pesticide fees during FY2006 pursuant to P.L. 108-199.
2007	Revised Continuing Appropriations Resolution for FY2007 (P.L. 110-5, H.J.Res. 20), enacted Feb. 15,2007[a]	Provided continued authorization for the collection of pesticide fees during FY2007 pursuant to P.L. 108-199, by providing funding under the authority, conditions, and limitations provided in the applicable appropriations Act for FY2006 (P.L. 109-54).
	S. 1983 introduced August 2, 2007, basis for PRIA 2 became P.L. 110-94, enacted Oct. 9, 2007	Reauthorizes, with some modification, PRIA 1 as enacted Jan. 23, 2004, in the FY2004 Consolidated Appropriations Bill (P.L. 108-199; Div. G, Title V).
	FY2008 Consolidated Appropriations Act (P.L. 110-161; Division F, Title II), enacted Dec. 26, 2007	Provided continued authorization for the collection of pesticide fees during FY2008 pursuant to P.L. 110-94.

Table 1. (Continued)

Year	Legislation/Regulation	Pesticide Fee Authority/Action
2008	Consolidated Security, Disaster Assistance, and Continuing Appropriations Act, 2009 (P.L. 110-329), Division A, Continuing Appropriations Resolution, 2009, enacted September 30, 2008	Generally extended funding for accounts in the Interior, Environment, and Related Agencies Appropriations bill at the amounts provided in the FY2008 regular appropriations law. Funds are available under the terms and conditions provided in that FY2008 law, except where otherwise specified.
2009	Omnibus Appropriations Act, 2009 (P.L. 111-8; Division E, Title II), enacted March 11, 2009	Provided continued authorization for the collection of pesticide fees during FY2009 pursuant to P.L. 110-94.
	Interior Department and Furthering Continuing Appropriations for 2010 (P.L. 111-88; Division A is the Department of the Interior, Environment, and Related Agencies Appropriations Act, 2010), enacted October 30, 2009.	Provided continued authorization for the collection of pesticide fees during FY2010 pursuant to P.L. 110-94.

Source: Prepared by the Congressional Research Service from the relevant laws and *Federal Register* notices.

a. During the first session of the 109th Congress, the House and Senate Appropriations Committees reorganized their subcommittees, including placing EPA's appropriation under the Interior subcommittee after eliminating the VA-HUD and Independent Agencies subcommittee.

Congress amended FIFRA in 1996 (FQPA; P.L. 104-70), extending EPA's authority to collect the annual maintenance fees through FY200 1. FQPA also expanded the authority under FFDCA to include the use of fees for purposes of reevaluating "old" tolerances (tolerance reassessment). FQPA requires EPA to ensure "reasonable certainty" of "no harm," to analyze aggregate and cumulative effects of pesticides, and to apply safety factors for children. The new requirements introduced a host of responsibilities for EPA, particularly when establishing new tolerances and reassessing old tolerances.[48] After its expiration September 30, 2001, the statutory authority for maintenance fees was extended in annual EPA appropriations bills prior to the enactment of the PRIA provisions.[49]

Other Pesticide Fee Authority

In May 1988, prior to the 1988 FIFRA amendments, EPA promulgated a final pesticide registration fee regulation,[50] citing the authority of the Independent Offices Appropriation Act (IOAA) of 1952 (31 U.S.C. § 9701). Intended to defray increasing administrative costs of pesticide registration reviews, the final rule included a prescribed schedule of fees to be submitted with each application for registration, amended registration, or experimental use permit. Registration fees were to be deposited in the U.S. Treasury and not directly available to EPA. The regulation was challenged in court by the Chemical Specialties Manufacturers' Association,[51] and the collection of registration fees under this authority was temporarily suspended through FY1 997 by the 1988 amendments to FIFRA (§ 4(i)(6)). Collecting registration fees under this authority continued to be prohibited through FY2001 by the 1996 FIFRA/FFDCA amendments (FQPA) and, subsequently, by annual appropriations bills from FY2002 through the FY2004 Continuing Resolution.[52]

Proposed Pesticide Fee Authority Modifications

In June 1999, EPA proposed a rule restructuring tolerance fees[53] in an effort to cover the cost of establishing initial tolerances and tolerance reassessments, including tolerance activities for "other" ingredients (namely, inert ingredients[54]). EPA proposed as much as a tenfold increase and the retroactive payment of fees for tolerance petitions submitted and reassessments initiated after FQPA was enacted in August 1996. Industry groups generally opposed the proposal. According to comments submitted to EPA, several industry groups disagreed and were concerned with, among other issues, EPA's interpretation that the statute provided authority to collect 100% of the cost of tolerance reassessment using fees. These groups also generally opposed EPA's justification for the tenfold increase in fees, the imposition of fees retroactively, and the potential effects of imposing fees for inert ingredients.[55]

The 106th Congress prohibited promulgation of the tolerance fee rule in EPA's FY2000 appropriations (P.L. 106-74, § 432). The 107th Congress considered approaches to revise the overall fees structure for pesticide programs and incorporated one approach in a manager's amendment to the Senate version of the 2002 farm bill (S. 1731). The conference substitute deleted the fee provisions and was not included in the final Farm Security and Rural Investment Act of 2002 (P.L. 107-171). In the conference report accompanying the final bill (H.Rept. 107- 424, p. 666), the managers "strongly encouraged" EPA to withdraw its proposed tolerance fee rule and to instead work with the appropriate committees for a solution. Similar proposals to increase tolerance fees, included in EPA's annual budget requests for FY200 1 through FY2004, have been prohibited each year by Congress in appropriations acts.[56] As discussed earlier in this report, the PRIA 1 provisions enacted in 2003 prohibited the collection of *any* tolerance fees beginning in FY2004 through FY2008, and PRIA 2 continues this prohibition through 2012.

Despite the PRIA prohibitions on additional pesticide fees beginning in FY2004, the current and previous two Administrations proposed increased pesticide fees above those provided under PRIA in the FY2005, FY2006, FY2007, FY2008, FY2009, and FY2010 budget requests for EPA. The 108th Congress rejected the President's FY2005 budget proposal to reinstate pesticide fees in the conference report on the Consolidated Appropriations Act for FY2005 (H.Rept. 108-792, p.1597). In the first session of the 109th Congress, language contained in the FY2005 supplemental appropriations for military funding enacted May 11, 2005 (P.L. 109-13, Sec. 6033), banned EPA from going forward with rulemaking for collecting pesticide tolerance fees as prohibited by PRIA.

The President's FY2006 budget request included collecting $46.0 million in pesticide fees in addition to those authorized under PRIA by proposing to again finalize the tolerance fee rule and to eliminate the prohibition on collecting registration fees originally codified in 1988.[57] The FY2006 appropriations bill for Interior, Environment, and Related Agencies (P.L. 109-54, H.Rept. 109-188), which includes EPA and was enacted August 2, 2005, did not reflect the Administration's additional anticipated pesticide fee revenues. The proposed fee changes in the Administration's requests would have required congressional approval through the enactment of legislation. In its report on the FY2006 appropriations, the House Appropriations Committee noted that no relevant legislation had been proposed and commented that EPA should not continue to spend time and resources proposing such actions in conflict with current authority (H.Rept. 109-80, p. 105-106).

The President's FY2007 request included an additional $56 million in the form of "anticipated" revenues (offsetting receipts) to be derived from changes to fees for pesticide registrations. Proposed changes included a new "Registration Review" fee, elimination of the prohibition on collecting tolerance fees, a new schedule and restructuring of the registration service fee program, and increasing the authorized limit on the annual amount of maintenance fees to be collected. The 110th Congress did not adopt the Administration's proposed additional pesticide fee revenues. The Revised Continuing Appropriations Resolution for FY2007 (P.L. 110-5, H.J.Res. 20), enacted on February 15, 2007, provided FY2007 appropriations for EPA and numerous other federal agencies at the same level as provided under the authority and conditions stipulated in the applicable appropriations acts for FY2006 (P.L. 109-54), unless otherwise specified in P.L. 110-5. P.L. 110-5 included no such specifications for EPA's pesticide program activities.

The FY2008 budget proposed modifications to the current pesticide fees structure to collect $66 million in anticipated revenues,[58] which included a new fee program to offset the cost of implementing EPA's recently initiated Registration Review Program, restructuring of existing registration service fees, reinstatement of tolerance fees, and increasing the annual limit on the amount of maintenance fees to be collected.[59] The Consolidated Appropriations Act for FY2008 (P.L. 110-161) enacted December 26, 2007, which included EPA's FY2008 appropriations in Title II of Division F, did not reflect the Administration's additional anticipated pesticide fee revenues. However, PRIA 2, enacted October 6, 2007, amended the authority for use of funds in the Reregistration and Expedited Processing Fund to include use for Registration Review, and included mandatory adjustments of the registration service fees—a 5% increase beginning with registration applications received October 1, 2008, through September 30, 2010.

The FY2009, FY2010, and FY2011 budgets contained similar proposed modifications and increases for pesticide fees, to allow for the collection of an additional $51 million, $48 million, and $42 million in fees for each fiscal year, respectively. The budgets proposed restructuring of existing registration service fees, reinstatement of tolerance fees, and an increase in the annual limit on the amount of maintenance fees to be collected. The FY2009 proposals were not reflected in the Omnibus Appropriations Act, 2009 (P.L. 111-8; Division E, Title II), enacted March 11, 2009, and the FY20 10 proposals were not included in the Interior Department and Furthering Continuing Appropriations for FY20 10 (P.L. 111-88; Division A is the Department of the Interior, Environment, and Related Agencies Appropriations Act, 2010), enacted October 30, 2009. The House and Senate Appropriations Committees had not reported their respective bills for the FY20 11 Interior, Environment, and Related Agencies appropriations, which fund EPA's programs and activities, at the time this CRS report was updated.

PESTICIDE PROGRAM FEE REVENUES AND APPROPRIATIONS

Historical appropriated funding and fee revenues for the pesticide program activities provide context for the discussion of fees imposed on pesticide registrants to supplement EPA- appropriated revenues. The two sections that follow provide more detailed information regarding pesticide fee revenues over time and funds appropriated for EPA pesticide program activities.

Revenues from Pesticide Fees

The amount of pesticides fees collected over the years has varied, depending on the statutory authority at the time. **Figure 1** below provides a graphic illustration of the amount of tolerance fees, registration fees (only collected for a short period during FY1 988), reregistration fees, maintenance fees, and registration service fees collected during FY1 985 through FY2009, before and after the enactment of PRIA in January 2004. The highest combined amount collected from the three fees for one year prior to the enactment of PRIA was an estimated $39.1 million in FY1 990, the peak year for collection of the one-time reregistration fees. The highest combined total amount of registration service fees and maintenance fees collected after enactment of PRIA was an estimated $41.6 million in FY2004.

Registration Service Fees (PRIA)

Registration applications received on or after March 23, 2004, were subject to the new service fees under PRIA 1. In its FY2009 PRIA implementation report,[60] EPA reported receiving a net total of $16.1 million in new "registration service" fees during FY2009. The net total reflects subtracting $1.0 million in the form of refunds for overpayments and withdrawals of applications. EPA reported expending $18.5 million of the $25.5 million available during the fiscal year, which included $9.4 million carried forward from FY2008. The remaining FY2009 balance of $7.0 million is being carried forward to FY2010.

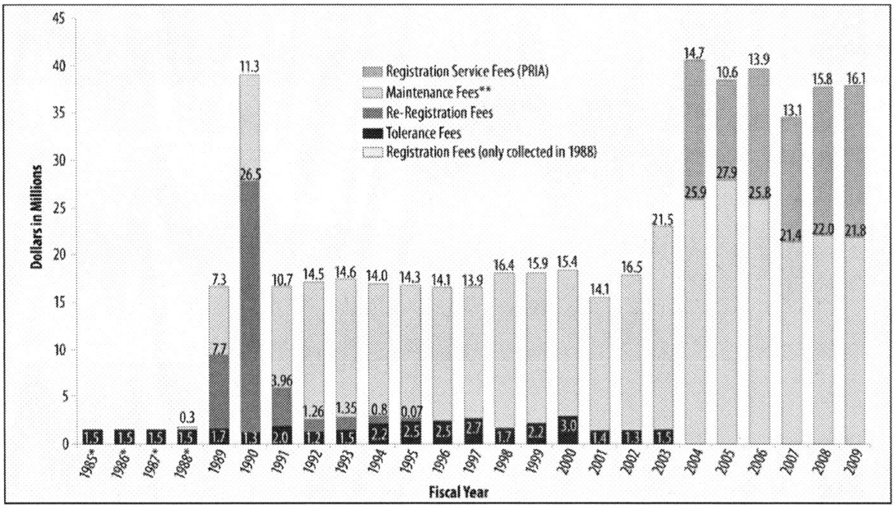

Source: Prepared by Congressional Research Service (CRS) with information from the U.S. EPA Office of Pesticide Programs.

* Tolerance fees for FY1 985-FY 1988 are based on the average number of petitions per year (8-12) and the average fee per petition ($150,000).

** Maintenance fees have been capped by legislation for each fiscal year: $14 million for FY1989-FY1997; $16 million for FY1998-FY2000; $14 million for FY2001; $17 million for FY2002; and $21.5 million for FY2003. PRIA 1 capped maintenance fees at $26 million for FY2004, $27 million for FY2005 and FY2006, $21 million for FY2007, and $15 million for FY2008. PRIA 2 extended the authority and set the annual statutory aggregate limit for maintenance fees at $22.0 million for FY2008 through FY2012.

Figure 1. EPA Pesticide Program Fee Revenues (Net Receipts), FY1 985-FY2009

The total FY2009 expenditure was a 9% increase above the FY2008 total expenditure of $17.2 million. The majority (51%) of the fee revenues expended in FY2009 was for payroll. The FY2009 payroll amount of $9.4 million (51% of total expenditures for that fiscal year) was a significant increase above the payroll amounts for FY2008 ($7.6 million; 44%), FY2007 ($7.1 million; 41%), FY2006 ($5.8 million; 54%), and FY2005 ($7.9 million; 71%). At $6.7 million (36% of fees expended), the proportion of fees used for contract expenditures in FY2009 was down slightly from the previous two fiscal years. The proportion of fees used for contract expenditures had been steadily increasing since FY2005: ($2.2 million (20%) in FY2005; $4.0 million (37%) in FY2006; $7.0 million (46%) in FY2007; and $7.2 million (42%) in FY2008.[61]

During FY2004, the initial year of collecting fees under PRIA, EPA indicated that it collected $14.7 million in new "registration service" and spent roughly $5.0 million. The remaining FY2004 balance of $9.7 million was carried forward to FY2005. Table 2 below presents EPA's reported expenditures of fees collected during each fiscal year FY2004-FY2009, including the distribution of expenditures by pesticide program activity.

Maintenance Fees

Annual maintenance fee amounts collected per registration are set in statute, dependent on the number of registrations held by a registrant. The fee amount is subject to adjustment by EPA, based on the annual aggregate limit, also established by statute. The initial 1988 authorization (P.L. 100-532) for maintenance fees set the annual aggregate at $14.0 million for the nine-year period from FY1989 to FY1997. The 1996 FQPA authorized collection of an additional $2 million (maximum aggregate of $16 million) per year for FY1998, FY1999, and FY2000, and returned to the original aggregate limit of $14 million in FY2001. The statutory authority for maintenance fees expired September 30, 2001, but was extended by Congress annually through appropriations legislation for FY2002, FY2003, and a portion of FY2004 to the enactment of PRIA. The annual aggregate limit was $17 million for FY2002 (P.L. 107-73), $21.5 million for FY2003 (P.L. 108-7), and $21.5 million for a portion of FY2004 through January 31, 2004, by Continuing Resolution P.L. 108-135.

PRIA 1, included in the Consolidated Appropriations Act of FY2004 (P.L. 108-199), extended the existing authority to collect maintenance fees through FY2008 at initially increasing, then declining, levels. PRIA set the annual statutory aggregate limit at $26 million for FY2004, $27 million for FY2005 and FY2006, $21 million for FY2007, and $15 million for FY2008. PRIA 2 extended the authority and set the annual statutory aggregate limit for maintenance fees at $22.0 million for each of the fiscal years FY2008 through FY20 12. Figure 1 above indicates that EPA generally collected the maximum aggregate limit as set by the statute in a given year. In the fiscal years following the enactment of PRIA, EPA collected $25.9 million in maintenance fees in FY2004 (EPA initiated collection of maintenance fees at the beginning of FY2004 under preexisting authority, prior to the reauthorization provisions included in PRIA 1), $27.9 million in FY2005, $25.8 million in FY2006, $21.4 million in FY2007, $22.0 million in FY2008, and $21.8 million in FY2009.

Table 2. EPA Expenditures from the Pesticide Registration Fund by Program Activity: FY2004-FY2009 (Dollars in Thousands)

Program Activity	FY2004	FY2005	FY2006	FY2007	FY2008	FY2009
Payroll	$2,535.3	$7,898.2	$5,819.8	$7,111.6	$7,556.4	$9,401.6
Contracts	$1,591.3	$2,228.8	$4,013.1	$6,979.5	$7,168.1	$6,733.3
Worker Protection	$430.0	$750.1	$750.0	$750.0	$2,250.0	$2,250.0
Other Expenses	$455.8	$274.3	$221.6	$302.7	$205.8	$140.6
Total Expenditures	$5,012.4	$11,151.4	$10,804.5	$15,143.8	$17,180.3	$18,525.5

Source: U.S. EPA, *Implementing the Pesticide Registration Improvement Act (PRIA) - Fiscal Year 2009*, February 26, 2010, http://www.epa.gov/pesticides/fees/.

Tolerance Fees

Historically, the annual tolerance fee collected from each applicant was based on the specific actions required to process a submitted application and varied depending on the number and type of petitions received by the agency in a given year. The amounts were adjusted over time based on an inflation calculation defined in statute.[62] For the 20 years prior to the enactment of PRIA, annual tolerance fees collected by EPA averaged about $1.8 million.

Reregistration and One-time Registration Fees

Reregistration fees varied considerably and were based, among other things, on whether the pesticide was an active ingredient registered for a major food or feed use or whether it was registered only for nonfood or nonfeed uses. The one-time active ingredient fee for reregistration ranged from $0 for a pesticide used exclusively for minor uses and for certain antimicrobial active ingredients to $150,000 for a major food or feed use active ingredient. By 1994, all authorized one-time reregistration fees had been collected, resulting in an estimated combined total of $31.64 million.

As discussed earlier, prior to the 1988 FIFRA amendments, EPA promulgated a final pesticide registration fee regulation,[63] under the authority of the Independent Offices Appropriation Act (IOAA) of 1952 (31 U.S.C. § 9701). EPA collected an estimated $300,000 in fees until the collection of registration fees under this authority was temporarily suspended through FY1 997 by the 1988 amendments to FIFRA (§ 4(i)(6)). Collection of these fees was suspended in subsequent years under the 1996 FIFRA/FFDCA amendments (FQPA) and later, annual appropriations (see discussion in "Other Pesticide Fee Authority").

Table 3. EPA Enacted Appropriations for Pesticide Program Activities, FY2004-FY2010 (Dollars in Millions)

Pesticide Program Activities by Appropriations Account	FY2004 P.L. 108-199	FY2005 P.L. 108-447	FY2006 P.L. 109-54	FY2007 P.L. 110-5	FY2008 P.L. 110-161	FY2009 P.L. 111-8	FY2010 P.L. 111-88
Environmental Programs and Management (EPM)							
Registration	$40.8	$39.2	$41.6	$40.4	NR	NR	NR
Reregistration	$51.7	$51.3	$57.5	$52.6	NR	NR	NR
Field Programs	$25.2	$24.4	$24.5	NR	NR	NR	NR
Science Policy & Biotech.	$1.7	$1.6	$1.7	NR	NR	NR	NR
EPM Subtotal	*$119.4*	*$116.5*	*$125.3*	*$119.5*	*$116.3*	*$116.1*	*$123.3*
Science & Technology (S&T)							
Registration	$2.3	$2.5	$2.4	NR	NR	NR	NR
Reregistration	$2.4	$2.5	$2.5	NR	NR	NR	NR
S&T Subtotal	*$4.7*	*$5.0*	*$4.9*	*$5.7*	*$5.8*	*$5.7*	*$6.6*
State and Tribal Assistance Grants (STAG)							
Implementation Grants	$13.0	$12.9	$12.9	$12.9	$12.8	$13.0	$13.5
Enforcement Grants	$19.8	$19.3	$18.6	$18.6	$18.4	$18.7	$18.7
STAG Subtotal	*$32.8*	*$32.2*	*$31.5*	*$31.5*	*$31.2*	*$31.7*	*$32.2*
Total	**$156.9**	**$153.7**	**$161.7**	**$156.7**	**$153.3**	**$153.4**	**$162.1**

Sources: Prepared by the Congressional Research Service based on information from House and Senate Appropriations Committees, conference reports and explanatory statements accompanying appropriations, and EPA's Congressional Budget Justifications for each of the fiscal years. Totals do not add due to rounding.

Note: NR indicates that a separate line item amount was not reported for that particular activity in a given fiscal year.

EPA Pesticide Program Appropriated Funds

Pesticide fee revenues are supplemental to appropriated funds provided for EPA's pesticide program activities. PRIA 1 and PRIA 2, in fact, included provisions to ensure that the fee revenues would not be offset by potential decreases in appropriations. As a condition of the statutes, authorization to assess registration services fees is suspended if congressional appropriations for specified functions conducted by the EPA Office of Pesticide Programs fall below a minimum amount relative to corresponding FY2002 congressional appropriations.[64] The minimum appropriations level has been met and the fee collection authority maintained for each of the subsequent fiscal years through FY2010.

In recent fiscal years, appropriated funding for EPA's pesticide program activities has been allocated within three of the eight EPA appropriations accounts: Science and Technology (S&T), Environmental Programs and Management (EPM), and State and Tribal Assistance Grants (STAG). Table 3 shows enacted appropriations for FY2003 through FY2010.

PESTICIDE REGISTRATION AND REREGISTRATION ACTIVITIES SINCE THE ENACTMENT OF PRIA

EPA uses registration service fees to supplement appropriations to develop improved registration review processes, hire new staff, and process registration applications under the deadlines identified in PRIA. The agency has used the maintenance fees to supplement appropriations primarily for reregistration (i.e., a reevaluation of pesticides registered prior to 1984) and tolerance review activities.

PRIA 1 had modified FIFRA with regard to completion deadlines for reregistration.[65] All reregistrations, other than those requiring tolerances for use on food, were to be completed no later than October 3, 2008. Reregistration of active ingredients that require tolerances or exemptions from tolerances were to be completed by August 3, 2006, as required by FFDCA (Section 408[q] [1] [C]) for tolerance reassessment. In anticipation of completing the pesticide reregistration and tolerance reassessment programs, EPA published a final rule for the "registration review" program in the *Federal Register* on August 9, 2006.[66] Through this registration review program, EPA plans to review pesticides registered as of October 2007 approximately every 15 years, consistent with FIFRA as amended. The agency began implementing the registration review program at the start of FY2007.

A key provision of PRIA 1 and PRIA 2, in conjunction with the increased fee revenues, is the requirement for EPA to identify reforms to its pesticide registration and registration review processes with the intent of reducing the agency's decision review times. EPA has reported implementation of a number of process improvements to monitor workload and ensure that PRIA due dates are being met. As required under PRIA, EPA released annual PRIA implementation reports describing these process improvements and other efforts and accomplishments. EPA released its inaugural report in March 2005 summarizing its first nine months of progress implementing the provisions of the PRIA from January 23, 2004, through September 30, 2004, and four subsequent reports were released in March 2006, March 2007, March 2008, March 2009, and February 2010.[67]

As discussed earlier in this report, EPA reported the completion of 7,530 decisions subject to PRIA between March 2004 (the effective date for PRIA implementation) and the end of FY2009. Among the 1,570 decisions completed during FY2009, 1,104 (70.3%) were for conventional pesticides, 342 (2 1.8%) for antimicrobials, and 124 (7.9%) for biopesticides.[68]

Registration Activities

At the end of FY2009 EPA reported that it had completed registration decisions (includes publications of import tolerances) for 19 new active ingredients, including 3 conventional pesticides (includes reduced risk pesticides), 15 biopesticides, and 1 antimicrobial.[69] EPA also reported that it approved 205 new uses of previously registered active ingredients by the end of FY2009. During the past six fiscal years (FY2004-FY2009), EPA completed decisions (either registration decisions or publications of import tolerances) for 141 new active ingredients, including 55 conventional pesticides (including reduced risk), 76 biopesticides, and 20 antimicrobials. During that same five-year period, the agency approved 1,374 new uses.[70]

EPA also reported the expenditure of $1.0 million in fees collected for worker protection program activities in FY2009, the same as FY2008 but an increase above the $750,000 during each fiscal year for FY2005, FY2006, and FY2007. PRIA 1 authorized the use of 1/17 of the registration fund (not less than $750,000 but not more than $1 million) for enhancing worker protection scientific and regulatory activities. PRIA 2 retained the same proportional distribution for worker protection, but increased the minimum to not less than $1 million.[71] Activities included continued interaction with stakeholder groups through the agency's Pesticide Program Dialogue Committee's Federal Advisory Committee, enhancement of safe practices and pesticide risk recognition training for workers and health-care providers to improve prevention and response, and expansion of poisoning recognition and occupational illness and injury surveillance.[72]

Also, as required under PRIA 2, $50,000 in fees was used for the Pesticide Safety Education Program, and $750,000 for partnerships grants during FY2009. Augmenting the allocated fee revenues with appropriated funds, EPA awarded $950,000 in partnerships grants to support joint public-private stewardship efforts to use and demonstrate the effectiveness of Integrated Pest Management (IPM) practices and technologies to reduce risks associated with the use pesticides.[73]

Among its efforts to enhance the registration process, has EPA reported that recommendations from several intra-agency workgroups led to the development of pesticide registration procedures for front-end processing and screening, waivers and refunds, funds management, improved intra- and interagency coordination, and enhancements to the internal registration tracking system. EPA also created a "Process Improvement" workgroup under the auspices of the Pesticide Program Dialogue Committee (an advisory group) to evaluate recommended process improvements in the registration program. The workgroup, which was further expanded, comprises representatives from individual registrant companies, pesticide trade associations, public interest groups, and agency staff, and it continues to address process improvement questions. Based in part on recommendations from the Committee, the agency continued to develop new process improvements during FY2009, and to refine those initiated during previous fiscal years. During FY2009, EPA concentrated outreach activities on communicating the difference between PRIA 1 and PRIA 2. EPA continues to focus on several areas identified by the stakeholder groups, including labeling consistency, communication of schedules, clarifications of certain application requirements and improved guidance materials, and electronic submissions.[74]

Reregistration/Tolerance Reassessment Activities

When it completes a review of a pesticide for reregistration or tolerance reassessment, EPA issues one of the following risk management decision documents: a Reregistration Eligibility Decision (RED), an Interim Reregistration Eligibility Decision (IRED), or a Tolerance Reassessment Progress and [Interim] Risk Management Decision (TRED).[75] Increased resources, and the adoption of integrated reregistration and tolerance reassessment process improvements under PRIA 1 have contributed to EPA more effectively meeting its statutory obligations.

By August 12, 2008, EPA completed reregistration decisions for 605 of the original 613 pesticide "cases,"[76] including 376 REDs and 229 canceled cases.[77] The eight remaining REDs were completed at the end of September 2008, meeting the reregistration statutory deadline. EPA reported[78] that it completed reassessment of 9,637 (99.1%) of the 9,721 preexisting tolerances by August 3, 2006 (the statutory deadline[79]). According to EPA, the reassessments resulted in the revocation of 3,200 food tolerances, the modification of 1,200 tolerances, and the retention of the remaining 5,237 tolerances. The remaining 84 tolerance reassessment decisions for five carbamate pesticides—aldicarb, oxamyl, carbaryl, formetanate and carbofuran—were completed in September 2007 as part of EPA's cumulative risk assessment for N-methyl carbamates pesticides.[80]

As a final step in the reregistration process, following the completion of an RED for a pesticide active ingredient, EPA must reregister the individual end-use products that contain a pesticide ingredient that has been declared eligible for reregistration. As of April 2009, EPA reported that it had completed decisions for 11,263 of the 22,122 pesticide end-use products subject to reregistration. Decisions are pending for 10,860 products. EPA completed 1,769 product reregistration decisions during FY2009, compared to 1,196 in FY2008 and 978 in FY2007. EPA's goal for FY20 10 is 1,500 decisions, and the agency expects to complete the last decisions in 2014.[81] EPA also reported that by the end of FY2009 it had completed product reregistration for 214 of the total 384 REDs, including 27 of the 31 organophosphates.

Registration Review[82]

The 1996 amendments to FIFRA (FQPA) included, among other things, a provision for the periodic review of pesticide registrations. Under section 3(g)(1)(A), Registration Review, "... The Administrator shall by regulation establish a procedure for accomplishing the periodic review of registrations. The goal of these regulations shall be a review of a pesticide's registration every 15 years." Such periodic review is necessary to determine if all registered pesticides continue to meet the statutory standard of no unreasonable adverse effects, taking into account changes in scientific capabilities for assessing risk, as well as changes in policies and pesticide use practices over time. As published in August 2006, EPA's registration review final rule replaces the agency's reregistration and tolerance reassessment programs, which are nearly completed.

According to EPA, there are 722 registration review cases, comprising 1,135 active ingredients.[83] The majority of these review cases are conventional pesticides (473 cases and 630 active ingredients), but also include antimicrobial pesticides, biochemical pesticides, and microbial pesticides. To meet its goal of reviewing each registered pesticide every 15 years, EPA anticipates it will require decisions on at least 45 cases (more than 70 pesticide active ingredients) annually. EPA initiated implementation of registration review in early FY2007, opening 25 registration review case dockets by the end of the fiscal year. Consistent with its published planned registration review schedule for FY2008 through FY2011, EPA opened 45 dockets in FY2008 and 70 dockets in FY2009.[84] The agency anticipates that it will continue at this annual pace (70 new dockets each fiscal year) through FY20 17, so that most of the currently registered pesticides will have dockets opened by the end of that fiscal year, with the exception of certain biopesticide dockets, which are expected in FY2018 through FY2020.[85] On June 8, 2010, EPA released an updated schedule for the registration review program that provides the timeline for opening dockets for the next four years, FY2010-FY2013, and summarizes updated information regarding the dockets opened FY2007-FY2009.[86] Among the changes to the schedule identified in the June 2010 release is the agency's decision to review all pesticides in two additional groups—triazines and soil fumigants—within the same time frame. The registration review of the soil fumigants group was moved forward from 2017 to 2013. EPA will update the registration review schedule at least once a year and possibly more frequently based on the agency's consideration of issues raised by the public, registrants, and other stakeholders.

CONCLUSION

EPA's timely completion of the statutory registration, reregistration, and tolerance assessment requirements for pesticides has been a concern for some Members of Congress, EPA, industry, and public interest groups. Historically, attempts to defray the increased costs of administering the pesticide program by modifying existing pesticide fee requirements through regulation and legislation have not been entirely successful.

The Pesticide Registration Improvement Renewal Act, or PRIA 2 (P.L. 110-94), enacted October 9, 2007, reauthorized and revised fee provisions enacted January 23, 2004 (P.L. 108-199, PRIA 1). PRIA 2 continued to address some of the key issues and concerns regarding EPA's pesticide registration reviews. Most notably, the provisions requiring specific decision process and schedule reforms, in conjunction with increased fee revenues, have led to more timely completion of certain registration applications and the completion of the reregistration reviews. Reforming the overall process has led to accelerated implementation of stricter FQPA standards and associated improvements in the safety of pesticides in the market. It has also contributed to the availability of new products, potentially safer and more effective, into the market sooner.

The prescriptive detailed schedules for the service fees included in PRIA 1 and PRIA 2, have been more commensurate with the specific EPA actions required than previous legislative provisions related to registration and tolerance fees, which were more generic. The implementation of the PRIA 1 schedules improved the efficiency in the overall process, leading to the completion of reregistration reviews and the reevaluation of older tolerances. PRIA 2's more prescriptive fee schedule, including increasing the number of fee categories from 90 to 140, is expected to result in further efficiencies of the processing of registration applications. Other PRIA 1 pesticide fee provisions retained in PRIA 2 also provided stability for resource planning purposes; stability had been lacking previously because of annual reauthorizations of maintenance fees and Administration budget proposals to modify fee authority.

EPA reported progress in developing process improvements and meeting shortened registration review deadlines during the PRIA 1 (PRIA became effective March 23, 2004) and continued to report refinements and improvements through the end of FY2009. Further improvement in the efficiency of the EPA's decision-making process under PRIA 2 is dependent largely on the agency's ability to continue to establish and effectively implement reforms while maintaining the protection of human health and the environment required by the statutes. To meet stricter statutory standards[87] and related "sound science" demands, EPA continues to develop and refine its scientific protocols and guidelines with input from stakeholders and the scientific community through various public forums.[88] However, as past experience has shown, this is a complex and time-consuming undertaking, affected by uncertainties and advances in technology that could enhance or inhibit further acceleration of the pesticide review process.

End Notes

[1] S. 1983 was passed by unanimous consent in the Senate on August 2, 2007, and by unanimous consent in the House on September 24, 2007.

[2] A registrant is defined as a person who has registered any pesticide pursuant to the provisions of FIFRA.

[3] Under § 3 3(k) of PRIA, EPA is required to publish an annual report describing actions taken under this section, and directed to include several elements. EPA released its inaugural progress report covering the period January 23, 2004, through September 30, 2004, in March 2005, and released subsequent fiscal reports on an annual basis thereafter. The EPA PRIA implementation report for FY2009 and previous annual reports for FY2004, FY2005, FY2006, FY2007, and FY2008 are available at http://www.epa.gov/pesticides/fees/.

[4] The 1988 amendments to FIFRA (P.L. 100-532) define "reregistration" as reevaluation of pesticides registered prior to 1984. The use of the term reregistration throughout this CRS report is as defined by the 1988 amendments.

[5] FFDCA §§ 408 and 409.

[6] FIFRA and FFDCA as amended in 1996 (FQPA; P.L. 104-170), "tolerance reassessments" are defined as those tolerances in existence as of August 1996.

[7] Section 408(o), as amended, the Pesticide Residue Amendment of 1954 (P.L. No. 518, 21 U.S.C. § 46(a)). The current authority resides in FFDCA § 408(m), per the 1996 amendments to FFDCA (FQPA).

[8] U.S. EPA, 64 *Federal Register* 31039-31050, June 9, 1999.

[9] EPA promulgated a rule for collecting registration fees under the authority of the Independent Offices Appropriation Act (IOAA) of 1952 (31 U.S.C. 9701). See Subpart U of CFR part 152, at 53 *Federal Register* 19108, May 26, 1988.

[10] *Chemical Specialties Manufacturers Association (now the Consumer Specialty Products Association) v. EPA*, No. 88-1525. D.D.C., July 25, 1988.

[11] Several industry groups disagreed and were concerned with EPA's interpretation that the statute provided authority to collect 100% of the cost of tolerance reassessment using fees. (EPA Docket # OPPT-301 151 and OPPT-301 151B.)

[12] Inert ingredients can be solvents or surfactants, and often compose the bulk of the pesticide product. Some inerts are known to be toxic; others are known to be harmless. EPA lists most in the category "non-food inert ingredients." See *http://www.epa.gov/opprd 001 /inerts/lists.html.*

[13] EPA's *FY2011 Annual Performance Plan and Congressional Justification*, and other related agency budget documents are available at http://www.epa.gov/ocfo/budget. Also see the multi-volume set of the President's *Budget of the United States Government, Fiscal Year 2011*, available at http://www.whitehouse.gov/omb/budget/Overview/.

[14] Relevant registration applications received between March 23, 2004, and September 30, 2007, were processed under PRIA 1 fees, decision review periods, and procedures.

[15] Enacted as Title V of Division G of the FY2004 Consolidated Appropriations Act.

[16] PRIA 1 (P.L. 108-199) removed the prohibition on "other fees" by amending FIFRA § 4(i)(6), replacing §§ 33 and 34 (7 U.S.C. § 36x and 136y) through 2010, and thus temporarily replacing registration fee authority codified in 1988 (Subpart U of CFR part 152).

[17] September 12, 2003, letter addressed to President George W. Bush, from a coalition of 30 organizations representing industry and public interests.

[18] See CropLife America's press release, October 11, 2007, at http://www.croplifeamerica.org/viewer.asp?pageid=220; see also Consumer Specialty Products (CSPA) and Natural Resources Defense Council (NRDC) joint press release, October 10, 2007, at http://www.cspa.org/.

[19] PRIA 1 had extended the statutory deadline for completing reregistrations for active ingredients that do not require tolerances to October 3, 2008; reregistration of active ingredients that require (food) tolerances or exemptions from tolerances were required to be completed by August 3, 2006, as mandated under FQPA (P.L. 104-170, Title IV, § 405).

[20] Under PRIA 1, annual maximum maintenance fees per registrant, and in aggregate, increased each year above the FY2003 levels for the first three years and declined in the final two years (P.L. 108-199, Division G, Title V, § 501(c)(1)(D) and (E)).

[21] Under the provisions of the 1988 amendments to FIFRA (P.L. 100-532), EPA calculates and adjusts the amount of annual maintenance fees collected per registrant, based on the number of registrants and the number of pesticide registrations, which is determined by the agency at the beginning of each fiscal year. The annual aggregate authorized under PRIA 1 increased from $21.5 million for FY2003 to $26 million for FY2004 and $27 million for FY2005 and FY2006; it declined to $21 million for FY2007 and $15 million for FY2008.

[22] Referred to as "Me-too" pesticides; see FIFRA § 4(k)(3)(i), "... the initial or amended registration of an end-use pesticide that, if registered as proposed, would be identical or substantially similar in composition and labeling to a currently-registered pesticide."

[23] See footnote 12, which defines the term "inert."

[24] Under PRIA 1, the registration fees schedule was per the September 17, 2003, *Congressional Record* (S11631 through S11633). EPA published the initial schedule of covered applications and registration service fees on March 17, 2004 (69 *Federal Register* 12771). In June 2005, EPA published a revised fee schedule (70 *Federal Register* 32327) based on a 5% increase in pesticide registration service fees, as authorized by PRIA (P.L. 108-199, Title V of Division G, § 33(b)(6)(B)), for applications received on or after October 1, 2005. Under PRIA 2 (P.L. 110-94, § (b)(6)), fees were increased another 5% effective October 1, 2008, in *Federal Register* August 5, 2008.

[25] 72 *Federal Register* 61465-61477, October 30, 2007.

[26] 73 *Federal Register* 45438–45450, August 5, 2008.

[27] U.S. EPA, "Pesticide Registration Improvement Renewal Act (PRIA 2) Fee Determination Decision Tree," at http://www.epa.gov/pesticides/fees/tool/index.htm.

[28] U.S. EPA, *Pesticide Registration Improvement Renewal Act (PRIA 2) Fee Category Interpretations*, July 9, 2009, http://www.epa.gov/pesticides/fees/PRIA2-Interpretation-table.pdf.

[29] 7 U.S.C. § 136w-8(b)(7)(F). During FY2004, EPA developed guidance for applying for waivers of the registration service fee under PRIA 1 and provided relevant information on a dedicated website. EPA also established formulae for reducing certain registration service fees. This guidance for registration service fee waivers and reductions, including recent updates, is available at http://www.epa.gov/pesticides/fees/questions/waivers.htm; information regarding the fee reduction formula is available at http://www.epa.gov/pesticides/fees/fee_reduction.htm.

[30] U.S. EPA, *Implementing the Pesticide Registration Improvement Act (PRIA) - Fiscal Year 2009*, February 26, 2010, http://www.epa.gov/pesticides/fees/.

[31] See previous fiscal years PRIA Implementation annual reports at http://www.epa.gov/pesticides/fees/.

[32] See footnote 30.

[33] 7 U.S.C. § 136w-8(c).

[34] FIFRA § 33(d)(2) Assessment of Fees (7 U.S.C. § 136w-8(d)(2)).

[35] FIFRA § 33(c)(4) Collections and Appropriations (7 U.S.C. § 136w-8(c)(4)).

[36] FY2004 (P.L. 108-199), FY2005 (P.L. 108-447), FY2006 (P.L. 109-54), FY2007 (P.L. 110-5), FY2008 (P.L. 110- 161), FY2009 (P.L. 111-8), and FY2010 (P.L. 111-88).

[37] 21 U.S.C. § 346a(m)(l). PRIA 1 prohibited collection of tolerance fees through FY2008; EPA published a notice suspending the collection of tolerance fees (69 *Federal Register* 12542, March 17, 2004).

[38] 7 U.S.C. § 136-1(i)(6).

[39] 7 U.S.C. § 136a-1(k)(1).

[40] The Food Quality Protection Act (FQPA) of 1996 amended FIFRA to add section 3(g), "Registration Review," with the goal of reviewing a pesticide's registration every 15 years.

[41] The revaluation of pesticides registered prior to 1984 as defined by the 1988 amendments to FIFRA (P.L. 100-532).

[42] The Food Quality Protection Act (FQPA) of 1996 amended FIFRA to add section 3(g), "Registration Review," with the goal of reviewing a pesticide's registration every 15 years.

[43] 71 *Federal Register* 45719.

[44] Sections 33(e), Reforms to reduce Decision Time Periods, and (f), Decision Review Time Periods.

[45] Each of the annual reports is available at http://www.epa.gov/pesticides/fees/.

[46] Pesticide Residue Amendment of 1954, P.L. No. 518, 21 U.S.C. § 346(a).

[47] This authority currently resides in FFDCA § 408(m) (1996 FQPA).

[48] See CRS Report 96-759, *Pesticide Legislation: Food Quality Protection Act of 1996 (P.L. 104-170)*, by Linda-Jo Schierow.

[49] The FY2001 statutory aggregate level of $14 million established by the 1988 FIFRA amendments was increased to $17 million in FY2002 (P.L. 107-73) and to $21.5 million in FY2003 (P.L. 108-10). The final Continuing Resolution for FY2004 (P.L. 108-13 5) extended the maintenance fees as authorized in FY2003 (see H.J.Res. 69, § 118).

[50] 40 CFR 152(u) and 40 CFR 172.

[51] *Chemical Specialties Manufacturers Association (now the Consumer Specialty Products Association) v. EPA*, No. 88-1525. D.D.C., July 25, 1988. The lawsuit has been held in abeyance since the passage of the 1988 FIFRA amendments.

[52] Appropriations bills for VA-HUD and Independent Agencies passed by the 107th Congress (P.L. 107-73) and the 108th Congress (P.L. 108-7; P.L. 108-135, Continuing Resolution for FY2004, expired January 31, 2004) contained similar prohibitive language.

[53] 64 *Federal Register* 31039-31050, June 9, 1999.

[54] The 1996 FQPA clarified that "inert" ingredients are covered by the definition of a pesticide chemical under FFDCA § 201(q)(1).

[55] EPA Docket # OPPT-301 151 and OPPT-30115 1B.

[56] Appropriations bills for VA-HUD and Independent Agencies passed by the 106th Congress (P.L. 106-3 77), the 107th Congress (P.L. 107-73), and the 108th Congress (P.L. 108-7, P.L. 108-135, FY2004 Continuing Resolution) contained similar prohibitive language.

[57] Executive Office of the President, Office of Management and Budget, *Major Savings and Reforms in the President's 2006 Budget*, available at *http://www.gpoaccess.gov/*us budget/browse.html.

[58] Executive Office of the President, Office of Management and Budget, *Major Savings and Reforms in the President's 2008 Budget*, availablc at *http://www.gpoaccess.gov/* usbudget/browse.html. See also *U.S. Environmental Protection Agency FY2008 Justification of Appropriation Estimates for the Committee on Appropriations*, available at http://www.epa.gov/ocfo/budget/.

[59] See http://www.epa.gov/oppsrrd1/registration_review/index.htm for more information regarding EPA's Registration Review Program.

[60] See EPA's *Implementing the Pesticide Registration Improvement Act (PRIA) - Fiscal Year 2009*, February 26, 2010, at http://www.epa.gov/pesticides/fees/; previous annual reports for FY2004, FY2005, FY2006, FY2007, and FY2008 are also available.

[61] See footnote 60.

[62] Tolerance fees could be adjusted annually, based on annual percentage changes in federal salaries (40 CFR 180.33[o]). The most recent adjustment in May of 2003 was an increase of

4.27%, based on the 2003 pay raise for General Federal Schedule (GS) employees in the Washington DC/Baltimore MD metropolitan area (68 FR 24370, May 7, 2003).

[63] 40 CFR 152(u) and 40 CFR 172.

[64] FIFRA § 33(d) Assessment of Fees (7 U.S.C. § 136w-8(d)(2)). Under PRIA 2, if congressional appropriations for a fiscal year fall below the amount allocated to OPP in FY2002 by more than 3%, authority to assess fees is suspended.

[65] 7 U.S.C. § 36a-1(g)(2)(A).

[66] 71 *Federal Register* 45719.

[67] The PRIA implementation annual reports are available at http://www.epa.gov/pesticides/fees/.

[68] EPA's *Implementing the Pesticide Registration Improvement Act (PRIA) - Fiscal Year 2009*, February 26, 2010, p. 9, http://www.epa.gov/pesticides/fees/.

[69] EPA's *Program Update-Registration*, presented to the Pesticide Program Dialogue Committee (PPDC) on October 7-8, 2008 (see http://www.epa.gov/pesticides/ppdc/).

[70] The number of new registrations and new uses were compiled by CRS with data from EPA's "Registration Activities In The Office of Pesticide Programs," October 15, 2009, http://www.epa.gov/opprd001/workplan/fy09finaldecisions.pdf; "New Biopesticide Active Ingredients - 2009," http://www.epa.gov/pesticides/biopesticides/ product_lists/new_ ai_2009.html; and "Antimicrobials Division 2009 Work Plan for FY2009," February 17, 2009, http://www.epa.gov/oppad001/pdf_files/adworkplan2009.pdf; as well as data provided by EPA's Office of Pesticide Programs in a written communication to CRS on May 19, 2010.

[71] 7 U.S.C. 138w-8(c)(3)(B).

[72] See Table I. "PRIA Funded Pesticide Safety Education and Worker Protection Activities in FY 2009," *Implementing the Pesticide Registration Improvement Act (PRIA) - Fiscal Year 2009*, February 26, 2010, pp. 30-35, http://www.epa.gov/pesticides/fees/. See also EPA's *PRIA Funded Pesticide Safety Education and Worker Protection Activities in FY 2008*, February 2009, for more detailed description of activities funded, http://www.epa.gov/pesticides/ fees/2008annual_report/pria-tbl-worker-prot-2-09.pdf.

[73] See "PRIA and Partnership Grants" in EPA's *Implementing the Pesticide Registration Improvement Act (PRIA) - Fiscal Year 2009*, February 26, 2010, pp. 8-9, http://www.epa.gov/pesticides/fees/. See also U.S. EPA, PRIA 2 Partnership Grants at http://www.epa.gov/pesp/pria2/index.html.

[74] See "Progress in Meeting Decision Times," pp. 9-14, and "Process Improvements in the Registration Program," pp. 15-25, in EPA's *Implementing the Pesticide Registration Improvement Act (PRIA) - Fiscal Year 2009*, February 26, 2010, *http://www.epa.gov* /pesticides/fees/.

[75] For more detailed explanation of these decision documents, see *http://www. epa.gov /pesticides/reregistration/* index.htm.

[76] Related pesticide active ingredients are grouped into cases; the 613 cases encompass approximately 1,150 pesticide active ingredients.

[77] EPA, http://www.epa.gov/oppsrrd1/reregistration/reregistration_facts.htm.

[78] EPA, *Accomplishments under the Food Quality Protection Act (FQPA)*, August 3, 2006, at http://www.epa.gov/ pesticides/regulating/laws/fqpa/fqpa_accomplishments.htm.

[79] The 1996 amendments to FIFRA and FFDCA included a schedule for completion of tolerance reviews: 33% by August 3, 1999; 66% by August 3, 2002; and 100% by August 3, 2006 (FQPA P.L. 104-170, Title IV, § 405).

[80] The Food Quality Protection Act requires EPA to assess the cumulative risks of pesticides that share a common mechanism of toxicity, or act the same way in the body. For more information on EPA's cumulative risk assessments, see *http://www.epa.gov* /oppsrrd1/ cumulative/index.htm.

[81] EPA, http://www.epa.gov/pesticides/reregistration/product-reregistration.htm.

[82] For a more detailed overview of EPA's registration review process see http://www.epa.gov/oppsrrd1/ registration_review/reg_review_process.htm.

[83] As of June 19, 2008; for updated information on EPA's schedule for opening dockets to begin pesticide registration reviews, see http://www.epa.gov/oppsrrd1/ registration_review/ explanation.htm.

[84] EPA, "Registration Review: Schedule for Beginning Reviews - 2009 to 2012," http://www.epa.gov/oppsrrd1/ registration _review/schedule.htm.

[85] See "Status of Registration Review," pp. 27-28, in EPA's *Implementing the Pesticide Registration Improvement Act (PRIA) - Fiscal Year 2009*, February 26, 2010, http://www.epa.gov/pesticides/fees/.

[86] EPA, "Registration Review: Schedule for Beginning Reviews - 2010 to 2013," http://www.epa.gov/oppsrrd1/ registration_review/schedule.htm.

[87] Stricter standards primarily refer to requirements introduced by FQPA in 1996 to perform more comprehensive risk assessment of pesticides, and consider aggregate exposure, cumulative effects from pesticides sharing a common mechanism of toxicity, possible increased susceptibility of vulnerable populations (particularly infants and children), and possible endocrine or estrogenic effects. (See CRS Report 96-759, *Pesticide Legislation: Food Quality Protection Act of 1996 (P.L. 104-170)*, by Linda-Jo Schierow).

[88] Examples of EPA advisory workgroups and committees for pesticide science and procedural issues are available at http://www.epa.gov/pesticides/science/committees.htm.

In: Pesticide Use in the U.S.: Law and Fee Issues ISBN: 978-1-61122-053-7
Editors: Regina M. Bianchi

Chapter 4

IMPLEMENTING THE PESTICIDE REGISTRATION IMPROVEMENT ACT (PRIA): FISCAL YEAR 2009

United States Environmental Protection Agency

SIXTH ANNUAL REPORT. REPORT RELEASE DATE

The Consolidated Appropriations Act of 2004 established a new system for registering pesticides, called the Pesticide Registration Improvement Act, or PRIA. The new section 33 of the Federal Insecticide, Fungicide, and Rodenticide Act (FIFRA), PRIA created a registration service fee system for applications for specified pesticide registration, amended registration, and associated tolerance actions, which set maximum residue levels for food and feed. Under PRIA, fees are charged for covered applications received on or after March 23, 2004, and for certain pending applications received before that date. The Environmental Protection Agency (EPA) is required to make a determination on the application within the decision times specified. The fee system was authorized until September 30, 2010. Due to the efforts of the PRIA Coalition of industry, trade associations, and public interest groups, PRIA was reauthorized on October 9, 2007 effective retroactively to October 1, 2007, the beginning of Fiscal Year 2008. The Pesticide Registration Improvement Renewal Act (PRIA 2) authorized the fee system to September 30, 2012.

Under FIFRA Section 33(k), EPA is required to publish an annual report describing actions taken under this section during the past fiscal year. The report must include several elements, including a review of the progress made in carrying out the Agency's obligations under the Act, a description of the staffing and resources associated with the review of and decision-making on applications, and a review of its progress in meeting the reregistration and tolerance reassessment timeline requirements. Additional PRIA 2 reporting requirements include information on electronic label review, a review of applications under section 3(c)(3)(B), and information on registration review that includes resource expenditures and recommendations for process improvements.

This sixth annual report covers Fiscal Year 2009 – October 1, 2008, through September 30, 2009, the second Fiscal Year under PRIA 2, and this report focuses on its continued implementation and impact. During FY 2009, the Agency improved its tracking systems, updated guidance, enhanced its application in-processing, and furthered the science of risk assessment.

PRIA 2 Enhancements in Application In-Processing

Previous annual reports (*2004, 2005, 2006, 2007,* and *2008*) described the steps the Agency undertook to implement PRIA. When PRIA 2 became effective, the Agency modified its in-processing procedures and processes. PRIA 2 increased the fee categories from 90 to 140 and changed payment procedures and how applications are screened upon receipt. Advancements were made in the screening process and additional guidance provided during FY 2009 based on experience during the first fiscal year of PRIA 2.

Front-End Processing and Screening Procedures - FY 2009

When PRIA 1 was implemented in 2004, the Agency established an intra-agency workgroup that interpreted the 90 PRIA 1 fee categories to help applicants and the Agency consistently place each application in the appropriate PRIA category. These PRIA registration categories reflected the types of applications the Agency may receive and for which Congress had established a specified fee and a time frame. The time frame, or decision review time, is the amount of time the Agency is expected to take to review the application and reach a regulatory decision. The intra-agency workgroup modified these interpretations for PRIA 2 and developed interpretations for the additional 50 fee categories. PRIA 2 revised or expanded the description of some PRIA 1 categories, requiring modifications in the Agency's interpretation of these categories. The PRIA 2 interpretations are available on the pesticides Web site on the *Fee Determination Decision Tree*. In 2009, these fee category interpretations were revised to provide additional guidance on the type of application that fell into each specific fee category. Stakeholders reviewed the proposed revisions to ensure that the guidance could be easily understood and applicants could accurately identify their fee category. The *fee category interpretations* were then posted on the internet. The Agency further modified its tracking system, Pesticide Registration Information System (PRISM), to improve its ability to track the new fee categories created by PRIA 2, to monitor fee activity, and to obtain the data needed to meet the additional PRIA 2 reporting requirements.

The Agency elected to invoice applicants instead of requiring payment at submission of an application under PRIA 1 because applicants were unfamiliar with the fee categories. Under this system, teams of EPA experts from the three registering divisions (conventional chemical, biopesticide, and antimicrobial pesticides) screened all incoming applications to determine whether they were subject to PRIA, assigned the application to a PRIA category if appropriate, and conducted a cursory screen of the application. The applicant was then invoiced for the appropriate amount with payment due within 45 days.

Certain provisions in PRIA 2 required the Agency to substantially modify this procedure. Under Section 33(b)(2)(D), the fee is due upon submission of the application. Section 33(b)(2)(F) directs the Agency to reject any application submitted without the required registration service fee. Consequently, the invoicing system was discontinued except when an additional payment is required. A portion of the fee (2 5%) is non-refundable once an application is submitted per Section 33(b)(2)(G). If any fee is unpaid 30 days after the fee is due, under Section 33(b)(2)(H), it is treated as a claim of the United States Government subject to subchapter II of chapter 37 of title 31, United States Code. A certification of payment is required with the application. Credit card or wire transfer payments can be made using the Treasury Department's *pay.gov* system, which provides an acknowledgement of payment. This acknowledgement of payment or a photocopy of a bank check serves as a certification of payment.

Since applicants have to identify the appropriate fee and pay it in advance of submitting an application (pre-payment) or upon submission to EPA, the Agency developed the *Fee Determination Decision Tree* and modified its tracking system, PRISM, to track payments and then match payments with applications, instead of with invoices. The Fee Determination Decision Tree is a tool that, through a series of questions and answers, allows pesticide registration applicants to identify an appropriate fee category and fee and was developed for inexperienced applicants. For experienced applicants, an "electronic short-cut" to the *fee interpretations* was also made available on the internet in FY 2009. The *interpretations* were also made available as a PDF table that included fees and time frames and can be printed and used as a hardcopy reference.

PRIA fees increased 5% effective October 1, 2008 and the Fee Determination Decision Tree and associated fee and payment guidance were revised by the effective date to reflect this increase. Payment information and a link to *pay.gov* for credit card and wire transfer payments are provided on the Decision Tree Web site. Once an application is received, the expert teams established under PRIA 1 screen the application and assign a PRIA 2 fee category. If the appropriate amount is not received, the Agency contacts the applicant and invoices the applicant for the unpaid portion, typically within 48-72 hours of receipt of an application.

The Agency treated the first quarter of FY 2008 as a transition period to provide enough time for applicants to become acquainted with the new payment procedures, and continued to send invoices requesting payment of the appropriate PRIA registration service fee if certification of payment was not received with the application. Beginning January 2, 2008, the Agency implemented a policy of not placing an application into the registration and review process if it did not contain certification of payment. The Agency would contact the registrant informing them that certification of payment was required together with the application. If certification of payment was not received within 14 days, the Agency would reject the application, and invoice the registrant for 25% of the appropriate fee. Nine applications were rejected in FY 2008 for failure to submit the appropriate fee while in FY 2009, only two applications were rejected for an unpaid fee.

21 Day Initial Content Screen

The cursory screen that the expert teams conducted under PRIA 1 had to be modified to implement Section 33(f)(4)(B), "Completeness of Application". This section directs the Agency, not later than 21 days after receiving an application and the required registration service fee, to conduct an initial screening of the contents of the application. In conducting this screen, the Agency must determine (1) whether the applicable registration service fee has been paid; or (2) at least 25 percent of the applicable registration service fee has been paid and the application contains a waiver or refund request for the outstanding amount and documentation establishing the basis for the waiver request; and (3) that the application contains all the necessary forms, data, and draft labeling, formatted in accordance with guidance published by the Agency. If the application fails the screen and can not be corrected by the applicant within the 21 day period, the Agency will reject the application.

The Agency phased in this screen. During a November 2007 PRIA 2 workshop, the Agency described its short term procedures for conducting this screen and its long term plans. Screening requirements were defined and a rejection process was developed by an intra-agency workgroup. Beginning January 2008, registering divisions assigned individuals or formed teams to conduct the screen of applications assigned to the division. A *screening worksheet* was developed, tested, and made publicly available on the pesticides Web site.

To ensure consistency across the pesticide program, the initial screen is now conducted by a contractor followed within 10 days by a review of the results by Agency employees. This review includes compliance with the Agency's formatting guidance, *PRN 86-5* Based on the experience gained during FY 2008, specifications were developed for this contractor support and it was phased in beginning with biopesticide applications on November 24, 2008, conventional herbicide applications on December 22, 2008, conventional insecticides on February 17, 2009, antimicrobials on March 9, 2009, conventional fungicides on March 23, 2009, and the remaining applications on April 6, 2009.

The Agency also phased in a more detailed review of Confidential Statements of Formula (CSF) (form *8570-4*) to identify unapproved inert ingredients during the 21 day screen. The 21 day screening contractor conducts a preliminary review to determine whether all of the inert ingredients are approved for the proposed uses using the lists of *approved inert ingredients* available on the Web. Staff in the Inert Ingredient Assessment

Branch (inertsbranch@epa.gov) confirm the status of any contractor identified unapproved inert ingredient. If an unapproved inert ingredient is identified and a request to approve the inert is not a component of the application package, the registrant is informed that data and a request to approve the inert ingredient needs to be submitted or the ingredient replaced with an approved one for the application to be further processed. The 21 day initial content review worksheet on the internet was modified with additional guidance on the Agency's review for unapproved inerts. In an analysis of CSFs submitted with conventional new product applications, the number of inert ingredient issues decreased from 26% to 7% as a result of this screen.

During FY 2009, 7 applications were rejected generally for missing or incomplete forms and data. In FY 2008, 14 were rejected.

FUNDS MANAGEMENT AND UTILIZATION

Section 33(c) of PRIA established the Pesticide Registration Fund. Congress established this fund in the Treasury of the United States to carry out the provisions of PRIA. All registration service fees received by EPA are deposited in this fund, and expenditures from the fund can cover the costs associated with review and decision-making for applications for which registration service fees have been paid. As of October 2007, fees are deposited into an account maintained by the U.S. Bank in St. Louis, Missouri, which informs the Agency when a payment is received. The later of date of payment or application receipt triggers the start of the PRIA decision review period, or time frame. The Agency has been informed of the receipt of a payment within an average of 7.2 days of receipt by the bank, and the Agency automatically sends an acknowledgment of payment to those applicants with an e-mail address on file.

The Agency encourages applicants to pay their fees by credit card or wire transfer using the Treasury Department's pay.gov system. These payments are more efficiently deposited with the U.S. Bank. In FY 2008, payments totaling $4,780,737 were made through pay.gov for 959 decisions. This represents 56% of the total number of actions for which payment was received. In FY 2009, payments totaling $5,804,462 were made through pay.gov for 1,150 decisions. This represents 65% of the total number of actions for which payment was received and an increase from 2008.

Under PRIA 1, EPA notified applicants when a payment was 45 days overdue for all PRIA fee categories except Fast Track applications (because of the short time frames for these actions). The notification gave the applicant 75 days to forward payment before the application was withdrawn by the Agency. In FY 2008 the Agency sent 59 such letters, resulting in 40 payments (totaling $400,312), 11 withdrawn applications, 4 fee waivers, and 4 determined to be secondary actions requiring no fee. These applications were received prior to PRIA 2's effective date. Applications received on or after October 1, 2008 are covered by PRIA 2's payment provisions: if payment is not received, the Agency rejects the application. In the case of a change in fee category to a higher fee during an in-depth review of an application, the Agency invoices the applicant for the difference and 75-day notices were sent if the payment was not received on the date due. In FY 2009, seven 75-day letters were sent resulting in one withdrawal and 6 payments totaling $56,726. In FY 2010, the Agency will no longer issue such 75-day notices and will reject an application if a payment is not received by the date due specified in the invoice.

Communications and Outreach

Communications and outreach efforts in FY 2009 focused on *PRIA 2 implementation* issues. Agency staff discussed PRIA 2 implementation during the Chemical Producers and Distributors Association Registration Workshop and annual meeting, meetings of CropLife America and the Consumer Specialty Products Association, with State and EPA Regional staff at the Pesticide Regulatory Education Program, and with the Armed Forces Pest Management Board. EPA provided updates on the status of PRIA actions received and summary statistics during meetings of the Agency's Federal Advisory Committee, the Pesticide Program Dialogue Committee (PPDC), PPDC Process Improvement Workgroup and meetings with the PRIA Coalition, composed of industry, trade associations, and public interest groups.

EPA also has quarterly meetings with the Biopesticide Industry Alliance to discuss PRIA and other common issues and with the United States Department of Agriculture (USDA) Inter-Regional Research Project Number (IR-4) program, and monthly teleconferences with USDA's Animal Health Inspection Service and the Food and Drug Administration on Plant Incorporated Protectants. The Agency has a standing bimonthly meeting with the Biocides Panel and Consumer Specialty Products Association to discuss issues such as those concerning antimicrobial applications and the due date extension process, international activities, the Primary Eye Irritation Pilot Program, Design for the Environment, factual statements, nanotechnology, guidance for evaluation of products against *Clostridium difficile* and 2009 Pandemic Influenza A H1N1. Presentations were made to the Association of American Pest Control Officials (AAPCO) and American Society for Healthcare Environmental Services (ASHES) on how antimicrobial pesticides are approved. A Webinar was held with Practice Greenhealth, a networking organization for the healthcare community to discuss the registration process for disinfectants and sanitizers with a focus on test methodologies.

Guidance on *fee category interpretations,* fee reductions resulting from *related applications,* and *refunds* were posted on the *PRIA 2 web site* and additional guidance was provided on the *21 day initial content screen review worksheet* and in the form of questions and answers. Through the PRIA Website, the public can submit questions regarding PRIA implementation. Questions are typically answered within 24 hours. Questions are also addressed by registration *Ombudsmen.* The Ombudsmen help applicants with issues related to identifying an application's fee and fee category, the implementation of PRIA 2, the registration process, and completing application forms.

Registration Program Workplans

The Agency's pesticide registering divisions continue to make their processes more transparent by providing additional information to the public on its Registering Pesticides Internet site such as workplans, schedules, and guidance. The multi-year *workplan for new conventional chemical actions* and new uses under PRIA is updated quarterly. These updates reflect new actions received under PRIA, actions completed, and changes to schedules. For a majority of the new chemical and new use actions listed, the time frame in which the Agency expects to complete its registration decision is shorter than that specified by PRIA. When possible, requests for new uses submitted by USDA's IR-4 program that are also being requested by registrants are merged to allow one risk assessment. Additional economies and time-savings were achieved where possible by folding new use assessments into assessments conducted for reregistration, and in the future, registration review. As they are registered, the Agency continues to post *risk assessments for new conventional pesticides* to aid registrants with future submissions. Human health and ecological risk assessments are attached to the new active ingredient fact sheets.

The *2009* and *2010* workplans for biopesticides are available on the Agency's pesticides website. The biopesticides workplan is updated at least once a quarter to reflect completed actions and changes to the schedule. *Biopesticide Registration Action Documents (BRAD),* posted on the Web for all new biopesticide decisions, give the basis for the Agency's decision, including a review of the studies submitted to support the registration.

The Antimicrobial's *FY 2009 workplan* for new antimicrobials and new antimicrobial uses was published and the FY 2010 workplan is anticipated in the near future.

FINANCIAL OVERVIEW

During Fiscal Year 2009, the Agency received $17.1 million in new registration service fees and, after subtracting $1.0 million in refunds (overpayments and withdrawals), net receipts were $16.1 million. A balance of $9.4 million was carried forward from FY 2008, including recoveries of prior year unpaid obligations. From this total of $25.5 million, the Agency spent approximately $18.5 million, carrying the remaining balance of $7.0 million forward to FY 2010. A balance is carried forward to fund personnel and contractor support for applications with multi-year time frames and for which some or most of the work is performed in the next fiscal year. Without a balance at the beginning of a fiscal year, staff would have to be reassigned from PRIA work until more fees were collected. This would disrupt the process and possibly result in missed PRIA deadlines. Spending increased by 8% in FY 2009, compared with FY 2008. The end of year remaining balance decreased by 26% in FY 2009 from FY 2008. As OPP staff time grew in order to meet PRIA deadlines, the major factor that increased spending in FY 2009 was the 24% increase in payroll charges.

Under Section 33(c), interest earned and added to the PRIA Registration Fund is available to the Agency for spending. Interest in FY 2009 totaled $7,432.

Payroll expenditures increased to $9.4 million in FY 2009 from $7.6 million spent in FY 2008. Expenditures on contracts decreased to approximately $6.7 million in FY 2009, compared with $7.1 million in FY

2008. Consequently, the balance between payroll and contract expenditures changed somewhat from 2008 to 2009 (with payroll at 51% of expenditures in FY 2009 compared with 44% in FY 2008, and contracts down to 36% in FY 2009 from 42% in FY 2008). In addition to funds from the PRIA Pesticide Registration Fund, the registration program spent about $37.7 M from appropriated funds.

Agency's FY 2004 through FY 2009 Expenditures from the Pesticide Registration Fund Expenditures (in thousands)

For	FY 2004	FY 2005	FY 2006	FY 2007	FY 2008	FY 2009
Payroll	$2,535.3	$7,898.2	$5,819.8	$7,111.6	$7,556.4	$9,401.6
Contracts	$1,591.3	$2,228.8	$4,013.1	$6,979.5	$7,168.1	$6,733.3
Worker Protection	$430.0	$750.1	$750.0	$750.0	$2,250.0	$2,250.0
Other Expenses	$455.8	$274.3	$221.6	$302.7	$205.8	$140.6
Total	**$5,012.5**	**$11,151.4**	**$10,804.5**	**$15,143.8**	**$17,180.3**	**$18,525.5**

As was the case in FY 2008, spending on mandated programs totaled $2.25 million in FY 2009 under PRIA 2. These mandates included worker protection ($1.0 million), partnership grants ($0.75 million), and the Pesticide Safety Education Program ($0.5 million). The percentage of expenditures going to the mandatory programs was 12% in FY 2009 compared to 13% in FY 2008 due to an increase in overall expenditures in FY 2009. The Agency also continued to invest in upgrading its information management systems to track compliance with the PRIA review time frames, to meet reporting requirements, and to implement PRIA 2 requirements. Other funds went primarily to pay for *Federal Register* printing costs associated with PRIA registrations.

Waivers of and Exemptions from Registration Service Fees

Section 33(b)(7) of PRIA authorizes the Agency to reduce or exempt the registration service fee under certain specified situations. The maximum that a fee can be reduced for small businesses with less than $10 million per year in global gross pesticide sales is 75% of the fee. A portion of all fees (25%) is nonrefundable. A 50% reduction in the fee may be granted for a small business with less than $60 million in annual global gross pesticide sales. The Agency's *guidance for small businesses on applying for a fee waiver* for requesting a reduction of a registration service fee is available on the Web. Section 33(b)(7) also provides an exemption from a registration service fee for applications from *Federal or State* agencies and for applications solely associated with a tolerance petition submitted in connection with the *Inter-Regional Project Number 4* and in the public interest.

In FY 2009, the Agency granted 306 fee waivers and exemptions and denied 6 of the 320 fee waiver/exemption requests received as shown in the following table. The remaining 8 were pending review at the end of the fiscal year.

Waiver Type	Received	Granted	Denied	Withdrawn
75% Ultra Business	201	189	6	0
50% Small Business	77	76	0	0
IR-4	31	30	0	0
Minor Use	1	1	0	0
Federal State	10	10	0	0
Total	**320**	**306**	**6**	**0**

The average number of days required to grant a fee waiver in FY 2009 (25 days) was consistent with the time required in FY 2008 (24 days). The average amount of time it took the Agency to deny a fee waiver/exemption is greater due to the increased time that the Agency took in an attempt to resolve the issues. The time to deny a fee waiver/exemption ranged from 17 to 54 days and the average in each quarter is shown below. The average number of days to deny a fee waiver during a fiscal year decreased from 42 days in FY 2008 to 37 days in FY 2009. There were no denials in the 2nd quarter FY 2009.

Average Number of Days to Process Fee Waivers in a Quarter, FY 2009

Quarter	To Grant	To Deny
1st Q	28	36
2nd Q	24	--
3rd Q	24	31
4th Q	24	44

The total fees waived and exempted in FY 2009 was $6.9 million, which was consistent with past fiscal years except for FY 2007 ($11 .4M). This amount may increase once the eight pending requests have been resolved. The amount reported for FY 2008 in the FY 2008 annual report was $7.85 million and it increased to $8.18 million during FY 2009 due to changes in fee category to one with a higher fee upon an in-depth review of the application and once the 16 requests pending at the end of FY 2008 were resolved. The majority of the FY 2009 fee waiver/exemption requests, 77%, were IR-4 exemptions, and approximately 13% were small business fee waivers. The total amount waived for small businesses decreased again from past fiscal years. An increased amount was exempted in FY 2009 for applications from federal and state agencies in comparison to FY 2008. The first minor use fee exemption was granted in FY 2009.

Amount in Fee Waivers and Exemptions by Fiscal Year of Receipt and Type (in $1,000)

Fiscal Year/ Type	Small Business	IR-4	Federal/ State Agencies	Minor Use Waiver or Exemptions	Total
FY 2004	$3,699	$2,745			$6,444
FY 2005	$3,006	$5,460	$15		$8,481
FY 2006	$1,497	$4,226	$40		$5,763
FY 2007	$2,162	$8,342	$924		$11,429
FY 2008	$1,247	$6,908	$28		$8,184
FY 2009	$879	$5,326	$471	$209	$6,885

Fee Reductions

Section 33(b)(8)(C) authorizes EPA to issue discretionary refunds, including instances where the Agency had completed portions of the review of an application before March 2004. For fees required for pending new active ingredients and for applications pending prior to March 2004 where the registrant has offered to pay the registration service fee voluntarily, the Agency applied this refund provision as a credit toward the registration application service fee. In past fiscal years, the amount of registration service fees that were reduced declined each year from $3.7 million in FY 2004 to approximately $3,500 in FY 2007. In FY 2008, no voluntary payments were received, while in FY 2009, one voluntary payment was received with a fee of $1,365.

Reregistration and Expedited Processing Fund

In FY 2009, the amount of money from the Reregistration and Expedited Processing Fund (maintenance fees or yearly registration renewal fees) used to carry out new inert ingredient reviews under section 4(k)(3) totaled $0.6 million. This supported 4.5 work years. An additional $2.4 million from this fund were used to process fast track amendments and new products.

During FY 2009, the Agency's obligations charged against the Reregistration and Expedited Processing Fund to offset the cost of the reregistration and registration review programs and other authorized pesticide programs were $24.6 million and 153.9 work years. The Fund has two types of receipts: fee collections and interest earned on investments. Of the $21.8 million in FY 2009 receipts, more than 99.9% were fee collections.

Appropriated funds are used in addition to Reregistration and Expedited Processing Fund dollars. In FY 2009, the Enacted Operating Plan included approximately $38.3 million in appropriated funds for reregistration and registration review program activities. This supported 208.1 work years and $11.0 M in contract support which included data reviews, systems maintenance and enhancements and other expenses. The unobligated balance in the Fund at the end of FY 2009 was $4.1 million, including recoveries of prior year unpaid obligations.

PRIA and Pesticide Worker Protection

Under FIFRA Section 33(c)(3)(B), EPA is authorized to use 1/17 of the amount of the Fund (but not less than $1 million) to enhance current scientific and regulatory activities related to worker protection and approximately, $500,000 in each fiscal year, 2008 through 2012, for funding of the Pesticide Safety Education Program (PSEP).

The Agency worked closely with worker safety stakeholders through the Agency's Federal Advisory Committee, the Pesticide Program Dialogue Committee (PPDC), to determine which activities to enhance with PRIA funds. Based on the advice of the PPDC, the Agency decided to develop enhancements within the following focus areas: Prevention - Safety Training, Response - Poisoning Recognition, Sound Decision Data, and Inform - Risk Management. *Table I* lists the activities funded and their accomplishments in FY 2009.

PRIA and Partnership Grants

When PRIA was reauthorized, an amount from the fees collected were set aside for partnership grants in Section 33(c)(3)(B)(ii): specifically $750,000 each for fiscal years 2008 and 2009 and $500,000 each for fiscal years 2010 through 2012. In 2008, EPA augmented these funds with appropriated funds and awarded approximately $1 million in grants to fund five projects that use Integrated Pest Management (IPM) approaches to reduce pesticide risk with the funds to be spent over a two year period. In FY 2009, EPA again augmented PRIA 2 funds with appropriated funds to award approximately $950K in grants and fund four projects. *Table II* provides a summary of this grant program's accomplishments with FY 2008 funds and lists the projects awarded with FY 2009 funds which began in late 2009 after a competitive selection process. Fiscal Year 2009 grants support the demonstration of innovative IPM practices and technologies, as well as outreach and education, in California, Florida, Wisconsin, Maryland, Pennsylvania, and Michigan. Approximately 50% of the FY 2009 funds support IPM approaches in urban communities and residences.

The FY 2010 Request for Proposals and PRIA 2 Partnership Grants competition is targeted for early February 2010. For FY 2010, EPA will award $500K in PRIA funds, and the solicitation for proposals will include projects to be funded by the Office of Science Advisor for approximately $400K. Additional information is available on the *PRIA 2 Partnership Grants* website.

PROGRESS IN MEETING DECISION TIMES

Number of PRIA Actions Completed in FY 2009

The Agency completed 1570 decisions subject to PRIA during the fiscal year, 107 (6.4%) fewer than the 1677 completed in FY 2008. Among the FY 2009 completed decisions, 342 (21.8% of total) were antimicrobial decisions, 124 (7.9%) biopesticides and 1104 (70.3%) conventional pesticide decisions. The decrease in the number of conventional actions completed (139) resulted in the overall decrease in actions completed as shown below. The number of antimicrobial actions completed has steadily increased, while the number of biopesticide actions varied from one fiscal year to another. An additional 167 decisions were withdrawn (24 antimicrobial, 14 biopesticides and 129 conventional), while in FY 2008, 156 decisions were withdrawn with 22 antimicrobial, 10 biopesticides and 124 conventional decisions withdrawn. The number withdrawn has increased yearly from 2007 to 2009.

Type of Pesticide	**Number Completed in Fiscal Year**			**Number Withdrawn in Fiscal Year**		
	2007	**2008**	**2009**	**2007**	**2008**	**2009**
Conventional	1189	1243	1104	77	124	129
Antimicrobial	308	336	342	35	22	24
Biopesticide	123	98	124	24	10	14
Total	**1620**	**1677**	**1570**	**136**	**156**	**167**

EPA completed 99.7% percent of these decisions on or before their PRIA or extended due date. In FY 2009, four actions missed their statutory due date by one or two days due to processing delays. Among the actions withdrawn, the Agency exceeded the PRIA due date for five decisions due to environmental risk issues and these actions were subsequently withdrawn. Another was withdrawn two days after the due date.

Table III titled "Number of PRIA Actions Completed in FY 2007, FY 2008 and 2009", summarizes the number of decisions completed by PRIA category and compares FY 2007, FY 2008, and FY 2009. FY 2007 was the last year of PRIA 1 and results from FY 2007 are shown for comparison to the first two years under PRIA 2. A summary of the actions completed in FY 2005 and 2006 are provided in the FY 2008 PRIA Annual Report.

Actions received under both PRIA 1 and PRIA 2 were completed in FY 2009, and both types of fee category codes are shown. Actions with a fee category with two digits are PRIA 1 actions (e.g., R01, A53) while PRIA 2 actions have a three digit fee category (e.g., R010, A530). In reviewing the table, certain factors need to be considered. An application can have more than one decision. The number of decisions depends on the number of product registrations in an application. If a tolerance petition is included in the application, the petition is also assigned a decision number to allow the Agency to track it and ensure that it is completed by the PRIA due date for the application. For instance, in FY 2009, one conventional first food use application package had six decisions, one for the tolerance petition and five for the five products associated with the package. Information on the number of active ingredients and uses registered each year are reported during a meeting of the Pesticide Program Dialogue Committee, a Federal Advisory Committee. Generally each application categorized as a Fast Track, Non-Fast

Track New Product, identical/substantially similar new product, new product, Non-Fast Track Amendment or label amendment submitted with data, contains a single product and is a single decision.

The number of actions (or decisions) completed each year increased steadily by over 20% per year from FY 2005 through FY 2007 under PRIA 1, increased by only 3.5% in FY 2008 and then decreased in FY 2009 by 6.4%. Because more actions were received in FY 2009 than in either FY 2008 or 2007 and fewer actions were completed in FY 2009, the number pending at the end of FY 2009 was greater than at the end of FY 2008. In comparing completions in FY 2009 with FY 2008, there were 33 fewer new active ingredient decisions, 105 fewer new use decisions and 92 fewer new product decisions. The number of new product decisions completed has consistently decreased from FY 2007. The number of amendments completed increased (58) along with other types of minor actions such as study waivers, protocol reviews and tolerance decisions (59) between FY 2008 and FY 2009. These differences were due primarily to differences in the number of conventional decisions completed.

The average decision time for each PRIA category, shown in Table III, is the number of days it took the Agency to complete a decision once the application was received and payment was made or a fee waiver or exemption was granted. The time frames mandated under PRIA 2 generally remained the same as the last fiscal year under PRIA 1 for similar fee categories. Exceptions were a reduction in the time frame for conventional reduced risk pesticides and an increase in the time frame for biochemical new products that required more than product chemistry data. For instance, the time frame for a reduced risk food use new conventional active ingredient decreased from 21 months to 18 months, while a non-reduced risk new conventional active ingredient's time frame remained 24 months. Under PRIA 1, time frames decreased from one fiscal year to another, e.g., for an R17 the time frame in FY 2006 was 22 months and it was reduced to 15 months for FY 2007 and FY 2008. A decision's time frame is based on the fiscal year in which the application was received. The date that decisions completed in FY 2009 were received ranged from October 2004 to 2009, resulting in decisions completed within one fee category with different mandatory time frames. Consequently, the average decision time or the number of days the Agency took to complete a decision in the table can not be directly compared to the PRIA time frames mandated for FY 2009 for many types of actions. Statutory time frames under PRIA 2 and for some identical/substantially similar and new products, however, have been somewhat consistent from one fiscal year to another.

For similar PRIA 1 and 2 fee categories, the average decision review time frames for PRIA 2 decisions were generally lower in FY 2009 than in FY 2008 except for a few categories of actions. The amount of time to complete actions for conventional new products, conventional label amendments requiring review within the Registration Division, and antimicrobial label amendments remained the same. These actions have short statutory time frames that remained consistent between PRIA 1 and 2. An increased amount of time in FY 2009 was required for decisions on antimicrobial products with unregistered sources or no letter of authorization and some biopesticide actions. The number of biopesticide actions completed was small and as observed in past annual reports, the number is too small to make adequate comparisons.

In the FY 2007 report, the average decision times for conventional reduced risk new food use active ingredients and new food uses were greater than those of non-reduced risk decisions. The number of reduced risk decisions was too small for the Agency to conduct an adequate analysis in 2007. In FY 2008 the average decision times for reduced risk new active ingredients were lower than those for non-reduced risk decisions and this trend continued in FY 2009, probably the result of reduced time frames for reduced risk actions under PRIA 2. Even though decision review time frames for conventional new use reduced risk decisions decreased, the average for these decisions completed in FY 2009 was greater than the statutory time frame because of due date extensions.

For those actions with consistent statutory time frames between PRIA 1 and PRIA 2, the average decision time review period for antimicrobial identical/substantially similar new products were greater than the statutory time frame. Among the PRIA 2 decisions, the average decision time review period for biochemical new active ingredients, and amendments, some conventional new product categories, certain antimicrobial protocols, and antimicrobial new products were greater than the statutory time frame.

Among the FY 2009 completions, due dates for 303 (19.3%) decisions were extended by mutual agreement of the applicant and the Agency. The percentage of decisions completed with due date extensions has increased each fiscal year. During FY 2006, FY 2007, and FY 2008, 11%, 13%, 18% of due dates, respectively, were extended. Extensions generally resulted from missing or deficient data or information. Due dates were extended for 17.5% of completed conventional decisions, while in the previous fiscal year, 14.9% % were extended. Twenty percent of antimicrobial and 33.9% of biopesticides were extended while in FY 2008, 22% and 48% respectively were extended.

Number of Completed Decisions with Due Date Extension Compared to Total Completed

Fee Category	FY2007		FY2008		FY 2009	
	Number due date extensions	Total Completed	Number due date extensions	Total Completed	Number due date extensions	Total Completed
Antimicrobial (A)	77	308	74	336	68	342
Biopesticide (B)	52	123	47	98	42	124
Conventional (R)	78	1189	185	1243	193	1104
Total Decisions	207	1620	306	1677	303	1570

The number of decisions with due date extensions was approximately the same in FY 2009 as in FY 2008, though the percentage of completed decisions with due date extensions increased in FY 2009 due to the decreased number completed. As discussed previously, a new use application package can have a number of decisions while a new product or amendment application package will have only one decision in the Agency's tracking system. Many new product and amendment applications, however, are dependent upon each other's data as described in the *primary/secondary guidance* and if there are data issues, the due dates for all of the affected applications will be extended. Consequently, an analysis of due date extensions using decisions can only indicate trends from one fiscal year to another. The 303 decisions with due date extensions were in the following general types of applications received by the Agency.

Number of Decisions with Due Date Extensions by Type of Decision

Fiscal Year	New Active Ingredient	New Uses	New Products	Amendments	Other (EUP, tolerances, protocols, etc.)	Total with Due Date Extensions
2007	28	18	121	35	5	207
2008	29	94	142	31	10	306
2009	17	93	123	52	18	303

Of the due date extensions in the different types of fee categories, new active ingredients, new uses, etc, new active ingredients continued to have the highest percentage of extended due dates. In FY 2008 and FY 2007 approximately 38% of the new active ingredient decisions completed had extended due dates and the rate was similar in FY 2009, 37%. Due dates were extended due to data deficiencies unique to each application. The percentage of new product decisions completed (both identical/substantial similar (fast track) and non-fast track new products) with due date extensions remained the same between FY 2008 and FY 2009 at approximately 17%, an increase over FY 2007's 14%. A common reason for these extensions was product chemistry data deficiencies. In the FY 2008 PRIA Annual Report, the Agency reported that the percentage of new use completions with due date extension increased substantially in FY 2008 (27%), compared to FY 2007 (7%) due to the time required to resolve risk issues. The percentage of new use completions with due date extensions again increased in FY 2009 to 39% and for the same reason. A smaller percentage of completed amendments 12% had due date extensions. This rate was higher than in FY 2008 (8.4%) and FY 2007 (9.5%).

Note: Appendix A contains a *list of all application subject to PRIA completed during FY 2009* and includes the decision times for each application. *(Microsoft Excel Viewer* is needed to view this file.)

Under Section 33(k)(2)(A)(iv), the Agency is to report the number of applications completed for identical or substantially similar applications under section 3(c)(3)(B), including the number of such applications completed within 90 days pursuant to that section. There are two types of identical or substantially similar applications, new products and label amendments that require no data review. The former have been called in the past "Fast Track New Products" while the label amendments are still called "Fast Track Amendments".

Identical or substantially similar new products (formerly "Fast Track New Products") are subject to registration service fees and have mandated decision review time frames under PRIA. With the passage of PRIA 2, identical or substantially similar products or Fast Track New Products were further subdivided into additional fee categories, some of which have time frames greater than three months. The number of identical or substantially similar products with a three month time frame (A530, B660, B7 10, and R300) completed in FY 2009 was 299 of which 245 were completed within 90 days and 277 were completed within their three month PRIA time frame and the remainder had due date extensions. An additional 74 identical or similar new products with time frames of greater than three months (A53 1, A532 and

R301) were completed (57 within the PRIA statutory time frame and 17 with due date extensions). In comparison to FY 2008, the number of decisions with a three month time frame completed in FY 2009 decreased (299 versus 358) while the number of decisions completed with longer time frames increased (74 versus 37). The percentage of three month time frame decisions completed within their statutory time frame was the same in FY 2009 (92.6%) as in FY 2008 (92.4%).

The time frame for “Fast Track Amendments”, label amendments that required no data review, remained 90 days under PRIA 2 and these amendments are not subject to registration service fees. In FY 2009, the Agency completed 2640 fast track amendment decisions or actions (unaudited results - antimicrobial 864, biopesticide 154, and conventional 1622) which had 3092 submissions (891, 209, and 1992, respectively). Each decision can have a number of submissions, each with a time frame of 90 days. Of these submissions, 2792 were completed within 90 days (884, 130, and 1778, respectively).

Section 33(k)(2)(E) directs the Agency to review its progress in meeting the timeline requirements for the review of antimicrobial pesticide products under section 3(h). The timeline requirement under section 3(h) for substantially similar or identical products is 90 days. Of the 49 decisions in fee category A530, 41 (84%) were completed within 90 days and 42 (86%) completed within the three month PRIA time frame. The remaining 7 had due date extensions. Of the 31 other substantially similar or identical products in fee categories A53 1 and A532, 19 were completed within their PRIA time frame of 4 months and the remaining 12 had due date extensions. Only 3 of these latter actions were completed within 90 days.

Regarding other new product decisions in fee categories A54, A540, A55, and A550, the section 3(h) time frame is 180 days with a goal of reducing the review time to 120 days. Of the 78 decisions in these fee categories, all met their PRIA due dates or extended due dates and 48 (62%) were completed within 120 days and 64 (82%) were completed within 180 days. The remaining 14 decisions had due date extensions.

Number of PRIA Applications Pending at the End of FY 2009

Table IV summarizes the pending registration applications (counted as decisions) in each of the PRIA categories. As of September 30, 2009, 1187 applications subject to PRIA were pending in the Agency's registration queue. Numbers pending at the end of FY 2008 and FY 2007 are shown for comparison – 1129 and 1207, respectively. The number of applications received in FY 2009 was greater than FY 2008 and FY 2007. As observed in the FY 2008 annual report, receipts in FY 2008 were approximately the same as in FY 2007. Because of the increase in receipts, decrease in the completions, and only a slight increase in the number of decisions withdrawn, the number pending at the end of FY 2009 increased by 58 over that of FY 2008.

The total number of applications (counted as "decisions") received in FY 2009 increased by 75 from FY 2008. The difference in receipts between FY 2008 and FY 2007 was eight decisions.

Totals	Difference in Number of Decisions Received between Fiscal Years		
	2007 and 2008	2008 and 2009	2007 and 2009
New Active Ingredients	14	9	23
New Uses	-7	-9	-16
Experimental Use Permits	-6	15	9
New Products	-148	25	-123
Amendments	77	16	93
Other (protocol reviews, tolerances, study waivers)	78	19	97
Total	**8**	**75**	**83**

Receipts in all categories of actions except new uses increased in FY 2009 over FY 2008. The numbers of new active ingredient, amendment and other types of decisions received have continued to increase since FY 2007. Among the other types of decisions are the actions in the new fee categories created by PRIA 2, and the difference in the number of other decisions received between FY 2009 and 2008 was due to an increase in the number of conventional protocols submitted for review. An increase in the number of conventional amendments (20) accounted for the general increase in amendments between FY 2009 and FY 2008. Conventional new product receipts have not rebounded after the substantial decrease in receipts (-124) experienced between FY 2007 and FY 2008. The increase in new product receipts between FY 2009 and 2008 was due to slight increases in the number of antimicrobial (13) and biopesticide (11) decisions received.

The number of antimicrobial decisions pending at the end of FY 2009 was approximately the same as FY 2008 (188 and 179, respectively), reflecting approximately level receipts (379 and 382), completions (342 and 336) and withdrawals (24 and 22). In considering decision type, the number of pending new uses (34 and 43) decreased while pending new products increased (82 and 68).

The pending number of biopesticide decisions was higher in FY 2009 (147 versus 127) due to increased receipts (161 versus 137). The difference in pending new product decisions accounted for the increased number of decisions pending in FY 2009 from FY 2008 (55 versus 33).

Among conventional pesticide decisions, the number pending at the end of FY 2009 increased from the end of FY 2008 (852 versus 823). The number of pending decisions was consistent between FY 2009 and FY 2008 for all types of actions except new active ingredient decisions indicating that the Agency was keeping up with conventional receipts. The number of new active ingredient decisions pending increased from 68 in FY 2008 to 96 in FY 2009 due to a decrease in the number completed (23 in FY 2009 versus 60 in FY 2008) and increase in receipts (54 in FY 2008 and 63 in FY 2009).

Pending Inert Ingredient Reviews at the End of FY 2009

FIFRA section 33(k)(2)(F) requires EPA to provide the number of inert ingredients (inerts) pending review by the Agency. In FY 2009, three new petitions for a food use inert were received as PRIA actions and an additional 45 new petitions were received as non-PRIA 2 actions. When PRIA was reauthorized, a request to approve an inert submitted with an application to register a conventional new product became subject to registration service fee requirements.

In FY 2009, 40 Final Tolerance Rules were published; one petition was denied; and five petitions were withdrawn due to deficiencies. At the end of FY 2009, there were 47 petitions in review, which included four petitions with deficiencies and 43 in various stages of review. All inert ingredient petitions are scheduled for review on the workplan in the order received. The Agency estimates that the average review time is 3-6 months for a polymer exemption petition and approximately 12 months (including data review, science assessment, decision document, and Final Rule) for a new inert petition. All new petitions are screened for deficiencies before being scheduled for review, and EPA works with prospective petitioners to discuss the reliability and adequacy of the data to meet the FQPA (Food Quality Protection Act) safety finding. In FY 2009, an additional 30 non-food use requests were granted and two non-food use requests were withdrawn or denied. Fourteen non-food requests were pending at the end of FY 2009.

The Inert Ingredient Assessment Branch (IIAB), consisting of seven employees at the end of FY 2009, reviews inert ingredient actions. When needed, IIAB staff are supplemented with staff in other pesticide regulatory groups, particularly to review inert ingredients associated with biopesticide and antimicrobial products.

PROCESS IMPROVEMENTS IN THE REGISTRATION PROGRAM

Section 3 3(e) of FIFRA directs EPA to identify and evaluate reforms to the pesticide registration process with the goal of reducing decision review times for pesticide registration applications. Section 33(k) directs the Agency to report its recommendations for process improvements in the handling of and streamlining of registration review. The Agency continued to make progress during the fiscal year in improving its operations. A number of steps were undertaken, internal and external, to explore, develop, and implement improvements in registration and registration review processes.

In improving processes, the Agency will not compromise the scientific quality of its assessments as a means toward reducing decision times. The Agency believes that the best means of gathering recommendations for process improvements is through the Federal Advisory Committee Act (FACA) process.

Pesticide Program Dialogue Committee PRIA Process Improvement Workgroup

The PRIA Process Improvement Workgroup was created in FY 2004 under the auspices of the Agency's Federal Advisory Committee, the Pesticide Program Dialogue Committee, to evaluate process improvements for the registration program. The workgroup is composed of members from pesticide registrant companies, pesticide trade associations, public interest groups, and

Agency staff. Meetings are open to the public and are held approximately twice a year. Reports of the April 20 and October 1, 2009, *PPDC PRIA Process Improvement Workgroup meetings* are posted on the internet.

Industry stakeholders and Agency staff identified many areas for improvement in the registration and registration review process, including labeling consistency, communication of schedules, use of electronic tools, application and submission guidance, and efficiencies in product reregistration and registration review. Many of the process improvements implemented by the Agency have addressed these issues. The Agency continues to work with all stakeholders to evaluate potential improvements to the registration and registration review processes. During Workgroup meetings, stakeholders present their priorities for process improvement and the Agency discusses the status of its improvement projects; previews new tools and proposed changes in procedures and processes; presents analyses of specific processes; and reports on its successes. Future projects and efforts are identified through a dialogue between the Agency and stakeholders.

Electronic Submission and Document Retention

The Agency is conducting a number of efforts to use information technology to improve the efficiency of the pesticide registration program and reduce the paperwork burden on both the Agency and the public.

In July 2008, EPA's Office of Pesticide Programs announced it would receive pesticide submission packages in electronic form or e-Registration submissions following a pilot project conducted in FY 2007. The Agency published a Federal Register Notice and provided *guidance* on the Web to broadcast this initiative. The types of applications currently being accepted electronically are Section 3 New Applications, Section 3 Amendments, Experimental Use Permits, Petitions for Tolerances, and applications for Supplemental Distributor Products. The Agency also established an e-Submission Help Desk in May 2008 to assist applicants with their questions about formatting their e-submission and to provide step-by-step direction to ensure the validity of the submission.

The e-Submission Module of the Agency's tracking system, Pesticide Registration Information System (PRISM) supports the processing of the documentation required for pesticide applications. Traditionally, this paperwork has been submitted in hardcopy form. The E-Submission initiative helps EPA move toward a more paperless environment. The information exchange from industry to EPA is based on a harmonized XML schema adopted from Canada's Pest Management Regulatory Agency (PMRA). This harmonization ensures that a submission package submitted to one participating regulatory agency can likewise be submitted to any of the other participating agencies, thus increasing standardization and decreasing the burden on pesticide applicants. Once the package is received by EPA, its contents are parsed and validated, thereby promoting data quality. The data submitted are then used to pre-populate data entry screens in an effort to save processing time and decrease the burden on EPA. Finally, the e-Submission module is fully integrated with PRISM's core data repository for registration information and its document management repository. When the incoming package has been processed within e-Submission, the data and documents are seamlessly blended into other PRISM components (Document Management Workflow) for processing within the pesticide programs. PRISM was enhanced to accept electronic registration (e-Registration) documents to make these documents available on-line at any time to the multiple users simultaneously processing registration actions. E-Submission/e-Registration will improve processing times, data quality and completeness; reduce data entry and the number of data entry iterations; and improve document management. In addition, EPA is actively working with the Organization for Economic Cooperation and Development (OECD) to develop a common, globally accepted standard for the transport of electronic data to the various international regulatory authorities. By conforming to a single standard, registrants will have to produce only one submission package for submission to multiple countries reducing time and resources required for multinational submissions.

Number of FY 2009 e-Submissions Compared to Paper Submissions

Quarter	Number e-Submission Packages	Number of Documents	% of all e-Submissions rejected	Number Paper Submissions	% e-Submissions of Total
1st	95	1714	36%	743	11%
2nd	103	1339	41%	954	10 %
3rd	100	1161	42%	896	10 %
4th	88	963	32%	1041	8 %

For fiscal year 2009, OPP received 386 e-submission packages. There were a total of 5177 documents associated with these 386 packages out of a total of 4020 submissions.

To facilitate the 21-day initial content review, a Workflow Management tool was developed for the registration program and in FY 2009, it was modified to notify the 21-day screening contractor that e-Registration documents are available on line for their review. This feature allows the contractor and staff to work with all electronic documents on line.

As a result of scanning documents and storing e-Registration documents, a Documentum library of over 190,000 documents is available electronically, an increase of 40,000 documents from FY 2008. Documents stored in the library consist of studies, forms, letters, and labels.

Electronic Labels

Acknowledging the Agency's efforts in this area, under PRIA 2, FIFRA Section 33(k)(2), the Agency is required to report the number of label amendments reviewed using electronic means and make recommendations for electronic submission and review of labels, including process improvements to further enhance the procedures used in electronic label review. The Agency and stakeholders recognize that reviewing labels can be time consuming. Procedures for submitting *labels in electronic format* have been in place for eight years. Registrants must format e-labels as text pdfs, name the file following a specific syntax, and submit the e-label either a) as a stand-alone file on a CD-ROM along with a paper application or b) as part of an XML file on a CD-ROM submitted without paper. Agency staff can then use Adobe Acrobat software to a) compare the proposed label against a previous version to identify changes and b) mark any required corrections on the e-label. Revisions of e-labels can usually be done using e-mail to exchange revised pdfs.

In the past, the Agency has been able to count the number of e-labels submitted but not the total number of label actions received in the same time period (and thereby calculate the percentages of paper and e-labels). To enable these statistics to be calculated, the Agency has been modifying its tracking system over the past two years to track electronic label submissions and to enable its staff to record when labels were reviewed electronically. This effort was completed at the beginning of July 2009. The statistics reported for FY

2009 are an estimate based on actual data from the fourth quarter, which were then multiplied by four to extrapolate to the entire year. Since label submissions are not necessarily constant throughout the year, these results are estimates and statistics reported in subsequent fiscal years can not be directly compared to these results.

FY09 Labels Submitted*

Type of Product	# labels submitted	% electronic labels
Antimicrobial	2,368	7 %
Biopesticide	540	7 %
Conventional	9,444	15 %
Unassigned	124	10 %
Total	**12,476**	**13 %**

* Data from the fourth quarter of FY09 was multiplied by four to reflect a full year of submissions.

FY09 Labels Reviewed*

			electronic label reviews	
Type of Product	**# all labels reviewed**	**# e-labels reviewed**	**% of all labels**	**% of e-labels**
Antimicrobial	2,664	160	5 %	80 %
Biopesticide	340	20	5 %	80 %
Conventional	8,156	780	6 %	58 %
Unassigned	48	8	8 %	50 %
Total	**11,208**	**968**	**5 %**	**62 %**

* Data from the fourth quarter of FY09 was multiplied by four to reflect a full fiscal year.

Based on these estimations:

1. Of approximately 12,500 label actions submitted to EPA in FY 2009, 13% included an electronic label.
2. Of approximately 11,200 label actions completed by EPA in FY 2009, 9% included an electronic label.
3. Of the label actions completed by EPA in FY 2009 that included an electronic label, 62% were reviewed electronically.

When considering the type of pesticide product, almost three quarters of the label actions submitted were for conventional products. Twice the percentage of conventional label actions was in electronic format as compared to antimicrobial and biopesticide label actions. However, of the electronic labels received, a higher percentage was reviewed electronically by antimicrobial and biopesticide staff than by conventional product staff.

Note: The two tables should not be compared to each other since they count different labels. Labels are usually not reviewed until any studies submitted with an action have been reviewed. Therefore, labels submitted in FY 2009 may not be reviewed until FY 2010. Conversely, labels reviewed in FY 2009 may have been submitted in earlier years.

Adoption of the e-label procedures by both registrants and EPA staff has been increasing slowly over the years. Registrant procedures for creating and submitting *e-labels* are detailed on EPA's website and have been presented and will continue to be presented in numerous industry meetings. All EPA label review staff were trained in the review of e-labels in 2007. Notification review staffs were trained in January 2010. Additional training classes for both registrants and Agency staff and one-on-one coaching are planned during 2010.

As a next step in improving electronic label submission and review, EPA prepared a comprehensive list of requirements for electronic labels (e-Label) in FY 2008. The objective of this effort is to improve the process of reviewing labels electronically and provide a tool for applicants to use in developing their labels. The general premise of the e-label is to identify all label content in a structured manner so that individual label elements can be processed with automated systems. Structured label content is expected to still further improve the efficiency of the label review process, allow for greater consistency of labeling content, and facilitate the distribution of product labeling via the Internet. Based on those requirements, the Agency is developing an XML schema definition that describes the technical specifications for the structured label content. Also under development is an application to load e-labels into the internal data system. The preliminary requirements for the "builder" application, to be used to create the needed e-label file, were gathered and the e-Label Loader application was developed in FY 2009. Once the other components of the structured e-label are finalized and implemented, the e-label builder and loader are expected to reduce the amount of time required for an applicant to prepare a label and for the Agency to review it.

Labeling Committee

The Agency formed a cross-program Labeling Committee in FY 2005 to address broad labeling issues and to oversee revisions to the *Label Review Manual.* A subgroup, the Label Review Manual Team, was formed to revise and continually update the Label Review Manual. During FY 2009 the Team completed its review and revisions to the Manual. All of the revised chapters have been posted to the Web; however, Chapter 13 will have some additional edits in the near future to reflect changes in storage and disposal regulations.

The Committee developed a Web site to communicate its activities and to address the *public's labeling policy questions* forwarded through the Web site's e-mail address (OPP _labeling _consistency@epa.gov)._The Committee received 105 questions during FY 2009 (a total of 312 questions since the site began). Answers to the majority of these questions were posted while some received a direct response. The Agency reorganized the question and answer portion of the Web site to provide more useful subject matter categories. The reorganized Web page was posted during FY 2009. At the suggestion of stakeholders, the Agency now flags new questions and answers with the word "New" in yellow for 30 days. The date answers were approved by the Agency is now placed next to the associated question. To address concerns that responses being posted on the Web site might be perceived as requiring changes to labels already on the market, the Committee placed a disclaimer of that intent at the beginning of the question and answer Web page.

The Committee from time to time publishes *issue papers* on its Web site. For example an issue paper on *Chemigation* (PDF, 6pp, 63.8kb) was made available for comment in December 2008. The site is also used to publish compact summations of selected policies that might otherwise be difficult for interested parties to locate. For example, the Agency's policy on *warranty and disclaimer statements* is posted, and an additional policy summation regarding certain label claims for cleaning products is likely in the near future.

Product Chemistry

An analysis of the reasons for PRIA due date extensions in FY 2009 again showed that in each of the pesticide registering divisions up to 60% of the due date extensions involved product chemistry issues (including *inert* issues). Many common errors are made on the Confidential Statement of Formula (CSF) form on which registration applicants list the contents of a product. The

Agency has developed a smart CSF form *(e-CSF)* for applicants' use that became available on-line in spring 2009. The smart form informs the applicant when a required portion of the form has not been completed or if the percent composition column does not add to 100%. The completed form can then be submitted in an electronic file format, which will reduce the Agency's reliance on paper and data entry required for the current paper-based process and make information readily available within the pesticide registration program. Improvements in the e-CSF builder have been suggested, such as identifying unapproved inerts and improving the "help" menu. These suggestions will be evaluated during FY 2010.

When the source of a conventional active ingredient is changed to a different manufacturer or source, the manufacturing or synthesis process may result in different impurities, which the Agency must assess for potential toxicological concerns. The Agency has a process in place involving the human health risk assessors to evaluate these impurities. This was due to an increase in the number impurities being noted associated with source changes and the additional or other impurities identified in the manufacturing (technical) product.

Process Improvements Implemented within the Pesticide Registration Program

The Agency's success in meeting due dates was a result of its continued monitoring of the status of PRIA decisions and identification of efficiency measures that conserve resources and time. Internal processes and tracking systems were modified to incorporate the 5% fee increase and to more effectively track and monitor study waiver requests and protocol submissions. All automated reports were then appropriately modified to monitor late applications and payments, and to report, for instance, the status of fee collections, rejections and due dates.

Agency staff continued to meet regularly to monitor workload and compliance with PRIA due dates and to resolve fee category and interpretation issues. For instance weekly staff meetings focus on the status of pending decisions, due date extensions, and refunds; identifying and resolving potential issues; resolving fee category questions; and coordinating schedules with science support organizations. During these meetings, staff identified PRIA 2 implementation issues and then discussed them with senior management during bi-weekly senior management meetings to ensure consistent implementation of PRIA across the pesticide program. Reasons for not granting, rejecting, or extending a due date are reviewed for consistency and to identify improvement opportunities such as incorporating additional reviews into the 21 day content screen. Senior managers continued to review justifications and make final decisions to extend or negotiate a PRIA due date and whether or not to issue a "PRIA Determination to Not-Grant" as under PRIA 1.

Staff reviewed policy decisions made during PRIA 1 for consistency with PRIA 2. This led to modifications in the questions and answers posted on the PRIA 2 web site and a detailed analysis of the manner in which discretionary refunds were given for applications dependent upon other applications which resulted in the *related applications guidance* posted on the internet. Inconsistencies in fee categories between A, B and R categories were identified particularly in how applications with an unapproved inert and requests to approve an inert were classified. This resulted in additional guidance on inert ingredients for the 21-day initial content review worksheet.

Pandemic 2009 H1N1 Influenza A Virus (Formerly Called Swine Flu)

The Agency published guidance to streamline the process for approving *H1N1 Influenza A Virus* claims on product labels. Typically, if registrants requested that this claim be placed on their labels, they would need to submit "virus specific" efficacy data under PRIA, a 120 day review process. The Agency worked with the Centers for Disease Control. By using the *disinfection hierarchy,* the Agency concluded that if a disinfectant product bears a label claim against human, avian, or swine influenza A virus, and efficacy data supported these claims, then a registrant may add Pandemic 2009 H1N1 Influenza A Virus to the product label via notification, a thirty day review process.

Science Review Improvements

OPP Science Policy Council

The Agency continued to improve the scientific basis of its review of and decision-making on applications. In June 2009 the Agency established the *OPP Science Policy Council.* The purpose of the council is to enhance the consistent use of the best available science in regulatory decisions and policies by providing a central forum that assists in identifying critical issues in pesticide safety, formulating solutions, and in transitioning new science, methodologies, and policies into the pesticide program. The Council focuses on cross-cutting science issues relevant across the program. The key functions of the Council are anticipating and proactively addressing emerging issues, providing research priorities, and transitioning new science policies and methods into its processes.

Over the next several years, the Agency will improve and transform its approach to pesticide risk management by enhancing its ability to use integrated approaches to testing and assessment in a manner consistent with the 2007 National Research Council (NRC) of the National Academy of Sciences report on Toxicology Testing in the 21st Century. An integrated approach will enhance the quality and efficiency of risk assessment and risk management decisions. A priority of the Council is to promote this transition.

Ecological Risk Assessments

The Agency continued to improve its review and communication of ecotoxicity studies through the following efforts: joint review/work sharing of study reviews with other countries; harmonization of ecotoxicity endpoints with other EPA programs; verification of drift reduction technologies; implementation and analysis of a global survey on pollinators; development of alternative approaches to estimate toxicity to aquatic life; harmonization of testing guidance with other countries; refinement of ecological risk assessments for pesticides with persistent, bioaccumulative, and toxic characteristics; and publication of peer-reviewed ecotoxicity values. Examples of these improvements include the following:

Joint reviews with OECD members: Working with the Organization for Economic Cooperation and Development (OECD), the Agency identified more efficient means to conduct joint reviews and work sharing, thus reducing review times and workload. The Agency also shared technical information with OECD countries and developed joint projects that will further improve the joint reviews of ecotoxicity studies. Examples of these joint projects include the harmonization of the terrestrial field dissipation guidance with OECD and the development of a crosswalk of ecoregions between North America and European countries. The Agency is also beginning to work with China in developing harmonized pesticide risk assessment approaches, conducting two training workshops on human health and ecological risk assessment in 2009.

OECD pollinator survey: The Agency also worked with the OECD to develop and implement a survey on pollinator testing programs, research, risk mitigation activities, and communication/outreach efforts, including the reporting of incidents. Responses were received from 17 countries and a final survey report was developed in October 2009. A high percentage of respondents indicated that declines in bee populations have been documented in their countries, and that declines have also been observed in other pollinator populations. Initial recommendations based on survey findings were proposed in November 2009, and included developing a mechanism to facilitate reporting of pollinator incidents among countries, coordination in the development of additional studies on toxicity to pollinators, sharing information on effective risk mitigation methods used in different countries, and mechanisms for global co-ordination of research efforts related to pollinator declines to facilitate research.

[Q]SARs: In the registration review program, the Agency has developed an approach for identifying critical ecological effects data gaps early in the risk assessment process. The Agency has also proposed and implemented scientifically appropriate methods to conduct risk assessments by relying on alternative [quantitative] structure-activity relationships ([Q]SARS) tools such as ECOSAR (Ecological Structure Activity Relationships) and ASTER (Assessment Tool for the Evaluation of Risk) to estimate the toxicity of pesticides, their degradates/metabolites, and potential contaminants to aquatic organisms when measured values are not available. Using these alternative techniques reduces uncertainties, the workload, and review times for the Agency, and reduces the testing burden on registrants. If additional data are still needed after using alternative techniques, EPA will request additional testing from the registrants.

KABAM: EPA completed its quality assurance review of its bioaccumulation model called KABAM (KOW (based) Aquatic BioAccumulation Model) to estimate potential bioaccumulation of hydrophobic organic pesticides in freshwater aquatic food webs and subsequent risks to mammals and birds via consumption of contaminated aquatic prey. This model can also be used to estimate pesticide concentrations in fish tissues. The model is composed of two parts: 1) a bioaccumulation model estimating pesticide concentrations in aquatic organisms and 2) a risk component translating exposure and toxicological effects of a pesticide into risk estimates for mammals and birds consuming contaminated aquatic prey. In the bioaccumulation component, pesticide tissue concentrations in aquatic organisms are calculated for different trophic levels of a food web through diet and respiration. In the risk component of KABAM, pesticide concentrations in aquatic organisms are used to estimate dose- and dietary-based exposures and associated risk quotients for mammals and birds consuming aquatic organisms. The methods used in the risk component of KABAM are consistent with the pesticide program's current modeling approach for assessing risks to terrestrial mammals and birds described in *USEPA 2004a* (Overview of the Ecological Risk Assessment Process in the Office of Pesticide Programs), as implemented in the T-REX model (version 1.4.1). More information concerning this model can be found at: http://www.epa.gov/oppefed1/models/water/kabam/kabam user guide appendix f.html.

Aquatic Life Benchmarks: In response to requests from FIFRA state lead agencies and state water quality agencies, EPA developed 78 additional "benchmark" values for pesticides that can be used to interpret monitoring data and to identify and prioritize sites for further monitoring. The benchmarks, which are based on the most sensitive aquatic toxicity data, are estimates of the concentrations below which pesticides are not expected to harm aquatic life. The Agency has made benchmark values for 148 pesticides available to the public by posting them on its *Aquatic Life Benchmark* Web site and has developed a public docket providing easier access to the full ecological risk assessments for these pesticides. The Aquatic Life Benchmarks have been used by federal agencies, states, and others in interpreting monitoring data and in planning future monitoring efforts. EPA plans to update the webpage and accompanying docket annually, and to add to the number of chemicals represented. Information concerning these benchmarks can be found at the following web site: *http://www.epa.gov/*oppfead1/cb/csbpage/updates/2007/aquatic-life.htm.

OPP/OW Harmonization of Aquatic Life Assessments: In response to concerns raised by states and other stakeholders, EPA's Office of Pesticide Programs (OPP) continued to work with the Office of Water (OW) to develop a harmonized approach for assessing aquatic toxicity data. In 2009, OPP and OW developed a Scoping Document that describes the Agency's efforts to develop a consistent and common set of effects characterization methods for both programs. This document describes the Agency's initial thinking on integrating the approaches used by OW and OPP for characterizing effects of pesticides to aquatic life. The technical work to develop these approaches began in 2009 with a meeting that included scientists from the Office of Research and Development (ORD), OPP, and OW. An extensive program of outreach to stakeholders is planned in 2010, beginning with six public meetings, starting in January 2010 at locations throughout the U.S. In these public meetings, EPA will solicit input from regional stakeholders regarding methods, tools, and approaches being developed and evaluated by OPP and OW, as well as additional sources of data and types of values used by states for protecting aquatic life. Additional information about this topic is available on the following web site: www.epa.gov/oppefed1/cwa fifra effects methodology/index.html

Drift Reduction Technologies: In FY 2009, the pesticide program continued to work with EPA's National Risk Management Research Laboratory (NRMRL) and Region 9 to identify and verify effective pesticide spray drift reduction technologies (DRTs). Under the Environmental and Sustainable Technology Evaluation (ESTE) program, EPA developed a draft verification protocol (DRT). The DRT testing protocol was adapted from standard test methods and regulatory methods used in other countries and describes the testing approach that will be used to generate high-quality, peer-reviewed data for DRTs, including test design and quality assurance aspects. Both low-speed and high-speed wind tunnel tests were completed this year using a reference nozzle and two test nozzles to evaluate the performance of the generic DRT testing protocol. By the spring of 2010, EPA plans to finalize this testing protocol based on the test results performed by EPA and stakeholders. As a next step, EPA intends to encourage equipment manufacturers to voluntarily use the protocol for testing their equipment. Additional information is available on the following web site: *http://www.epa.*gov/etop/etc atpsdt.html.

Pesticides with persistent, bioaccumulative, and toxic characteristics: Standard pesticide risk assessment methods do not address the unique characteristics associated with pesticides that exhibit persistent, bioaccumulative, and toxic (PBT) characteristics. In October 2008, EPA convened a Scientific Advisory Panel (SAP) to review issues and methods for ecological risk assessment of *PBT-like pesticides.* This SAP peer review was an initial step in a process for making refinements to the Agency's ecological risk assessment practices for addressing pesticides with PBT profiles. Over the next years, EPA will return to the SAP for additional review of specific methods and tools related to ecological risk assessment of pesticides with PBT characteristics. Additional information about this SAP meeting can be found on http://www.epa.gov/scipoly/sap/meetings/2008/102808_mtg.htm.

EPA is also leading an OECD "PBT Expert Group" for harmonizing approaches for addressing PBT risk assessment issues. Current participants represent 7 countries and 2 international institutions. This group will be developing risk assessment guidance for pesticides with PBT characteristics. In addition, EPA is actively participating in the review of chemicals nominated for annexation into the Long-Range Transboundary Air Pollution (LRTAP)/ Persistent Organic Pesticides (POPs) Convention and the Persistent Organic Pollutants Review Committee (POPRC) of the Stockholm Convention. Recent technical review meetings with EPA participation included the LRTAP/POPs meeting June 2009 in Bulgaria and the Stockholm POPRC in October 2009 in Switzerland.

Atrazine Monitoring Issues: In December 2007, EPA presented the Comprehensive Aquatic Systems Model (CASM) to the FIFRA Scientific Advisory Panel (SAP). CASM was developed to determine a level of concern (LOC) that compares atrazine concentration and exposure durations from monitoring data to effects measured in a series of microcosm and mesocosm studies which were used in the original atrazine (IRED) assessment. During the May 2009 FIFRA SAP meeting, EPA presented new evaluations of the applicability of the revised CASM atrazine model to freshwater atrazine risk assessment and gave reasons for not supporting further development and application of this model. A simpler alternative to the CASM-based approach, Plant Assemblage Toxicity Index (PATI), was presented for relating atrazine surface water exposures to the microcosm and mesocosm effects. Other issues that were presented and a summary of the meeting can be found at the following web site:_http://www.epa.gov/scipoly/sap/meetings/2009/may/051209minutes.pdf.

In FY 2009, the Agency continued its efforts to incorporate tools in its aquatic risk and exposure assessments that will enable the Agency to identify specific geographic locations where risks may occur. As part of this effort, the Agency acquired and developed data, including the national-level SSURGO (Soil Survey Geographic) soils data, updating its land use data with the 2007 NLCD (National Land Cover Database), and deriving hydrograph data sets from the national NHD (National Hydrography Dataset) – as well as improving its tools to provide more accurate and relevant information about the potential effects of pesticides in the environment. These data and tools, which are being used in the Agency's risk assessments, allow EPA to more quickly identify the landscapes and water bodies that are most vulnerable to pesticide impacts on drinking water sources and on aquatic species, including endangered species. For instance, the Atrazine assessment incorporates these technologies.

Human Health Risk Assessments

The Health Effects Division (HED) reorganized its registration and reregistration branches into 7 Risk Assessment Branches that perform both registration and registration review activities as discussed in the FY 2008 annual report. The reorganization has positively impacted the Division's work processes. Accountability and efficiency increased as a result of an individual overseeing the entire risk assessment process for a specific pesticide.

Regarding science review committees, the Residues of Concern Knowledgebase Subcommittee (ROCKS) is leading the application of predictive *Tox 21 tools* for metabolites, residues, and environmental degradation products. In calendar year 2009, the Dose Adequacy Review Team (DART) met seven times on six different chemicals to select doses for carcinogenic studies in rats and mice, and most recently, for immunotoxicity studies. The Cancer Assessment Review Committee (CARC) met four times on four different chemicals to establish cancer classifications on these chemicals, and the Toxicology Science Advisory Council (ToxSAC) met twenty-four times on twenty-six chemicals to determine end-points of concern.

Improvements in the Agency's pesticide human health exposure and risk assessments were focused in three major areas in FY 2009: 1) update and improve assessment methodologies and processes by incorporating recent scientific advancements and presenting these improvements to the FIFRA Scientific Advisory Panel (SAP) meetings for advice and comment; 2) advance assessment tools to increase the confidence of risk calculations; and

3) globally harmonize the tolerance (MRL) setting process. Significant progress was made in each area.

Worker Reentry Exposure and Risk Assessment: In December 2008, the SAP reviewed occupational post-application exposure monitoring studies conducted by the *Agricultural Reentry Task Force (ARTF)* and considered the Agency proposal for the duration of workday in exposure assessments. The SAP concluded that the studies were robust and useful for exposure assessment purposes and that the proposed use of individual studies to represent 'clusters' of similar occupational post-application reentry activities and the approach and inputs used – specifically for workday duration and for multi-day exposure assessments (i.e., short/intermediate-/ and long-term) were reasonable. The Agency will update its policy based on the Panel's advice and once implemented, better characterization of exposure and risk are expected.

Antimicrobial Exposure Assessment Task Force (AEATF): The *AEATF* was formed by a number of antimicrobial pesticide registrants to develop additional data to better represent actual exposure levels for a wider range of antimicrobial pesticide handler activities. The risk assessment process could be streamlined with a generic exposure database that can be used to assess new antimicrobials or new antimicrobial use requests. The AEATF is collecting data by monitoring dermal and inhalation exposures of applicators. In FY 2009, the field phase of the *mop and wipe studies* were completed. The aerosol study protocol was presented and accepted by the *Human Studies Review Board (HSRB)* in October 2009. Once studies are completed, the results will be reviewed by the HSRB.

Antimicrobial Exposure Joint Venture (AEJV): The AEJV is an industry group that has gathered in-home residential antimicrobial product use information. They are gathering information to answer the questions of when, where, how, and how often are antimicrobial products being used. These data will be reviewed and assessed by EPA for use in its risk assessments.

Pyrethroid Common Mechanism of Action: In June 2009, the SAP reviewed the Agency's evaluation of laboratory studies and other scientific information on the common mechanism of action of *pyrethroid pesticides* contained in a draft science policy document, "Proposed Common Mechanism Grouping for the Pyrethrins and Synthetic Pyrethroids". Synthetic pyrethroids are a class of insecticides structurally based on the pyrethrins, botanical insecticides extracted from *Chrysanthemum* *cinerariaefolium.* The potential exposure of the general public to these pesticides has increased over the past decade due primarily to the decreased residential use of organophosphates and *N*-methyl carbamates. The basis for a shared (i.e., common) mechanism of action for these compounds includes 1) shared structural characteristics and 2) shared ability to interact with the sodium (Na) channels. Two different neurotoxicity syndromes are expressed and the Panel agreed that these pesticides should be separated into two subgroups, Type I and II pyrethroids. After the policy is finalized, the naturally occurring pyrethrins and synthetic pyrethroids will be subject to cumulative risk assessments.

Integrative Testing and Assessment: The Agency initiated work on a long term vision for the application of *Integrative Testing and Assessment (IATA)* or computer-aided methods and integrated tools to better predict potential hazards and exposures. Activities have included a PPDC work group; a Web page to communicate with stakeholders on IATA issues; a computer model for predicting estrogenic activity; a pilot of the use of non-animal testing to address labeling requirements for pesticides; and development of a pesticide metabolism database (MetaPath). (Q)SAR-related activities that support IATA include a project to enhance existing (Q)SAR models with pesticide- specific data, development of a QSAR guidance document, testing the predictive performance of (Q)SAR models on pesticides, integrating traditional studies with QSAR data and other information, and participation in the development of the proposed rule on the use of (Q)SAR to address data requirements for selected pesticides. For MetaPath, the Agency is populating the database with rat, plant, and environmental degradation metabolism data. These efforts have involved collaborations and partnerships within the Agency and with the US Food and Drug Administration and Canada's Pest Management Regulatory Agency (PMRA) experts on QSAR and MetaPath.

F1 Generation Study: The Agency is developing improved animal study designs to more reliably and efficiently address risks to children through its proposed enhanced F1 generation reproductive toxicity study as part of the greater IATA effort. The new enhanced F1 generation study will better target and more efficiently address potential toxicity on the reproductive, neurological, and immunological systems with the use of fewer animals. It is scheduled to be presented to the SAP in 2010.

Non-animal Testing for Eye Irritation for Certain Antimicrobial Products with Cleaning Claims: The Draize rabbit eye test is used to determine eye hazards and the required hazard labeling for pesticide products. The Agency initiated a pilot project in May 2009 with interested applicants, designed to evaluate the effectiveness of a specific alternative non-animal testing approach, as a potential replacement for the Draize rabbit eye test for labeling antimicrobial products with cleaning claims. The proposed testing strategy uses three assays; the Bovine Corneal Opacity and Permeability test (BCOP), the EpiOcularTM model (EO), and the Cytosensor Microphysiometer assay (CM).

The objective of this *voluntary pilot project* is to evaluate and gain experience with certain non- animal testing methods (*i.e., ex vivo* and *in vitro*). Data submitted under this pilot will be used in labeling decisions if the results are deemed by the Agency to be adequate and appropriate to support such regulatory decisions. Eighteen months after the start of the pilot project, the data will be compiled and analyzed and the final assessment document will be presented to OPP's Science Policy Council for their recommendations on this non-animal approach.

Global Harmonization of MRLs: A NAFTA-harmonized approach to establishing tolerances developed several years ago is currently being used by Canada, Mexico, and the U.S. to establish harmonized tolerances, which serves to encourage trade and minimize trade irritants among the *NAFTA* partners. To encourage such harmonization on a global scale, an OECD workgroup with EPA's participation and responsible for harmonizing the MRL-setting process among the OECD member states, is extending and expanding upon the previous tolerance harmonization work of the NAFTA Workgroup.

Section 33(k)(2)(A) directs the Agency to report its recommendations for the allowance and use of summaries of acute toxicity studies. To date no acute toxicity summary has been submitted to the Agency for review. However, if submitted electronically, they can be used to create a Document Evaluation Record (DER) or the Agency's review. Currently, OECD template study reviews submitted electronically with new active ingredient submissions are being used as a skeleton for the Agency's DERs.

PRODUCT REREGISTRATION PROGRESS AND ACCOMPLISHMENTS

Overall Accomplishments

Product reregistration is EPA's program for implementing reregistration eligibility decisions by ensuring that required risk reduction measures are reflected on pesticide product labels. EPA has completed its review of the safety of pesticide active ingredients first registered before November 1984 through the reregistration program. The results of EPA's reviews are summarized in Reregistration Eligibility Decision (RED) documents available on the *Agency's Pesticide Reregistration Status* Web site. After the Agency completed a RED for a pesticide active ingredient and declared it eligible for reregistration, individual end-use products that contained the pesticide still were required to be reregistered.

As of the end of FY 2009, 22,122 pesticide products are subject to product reregistration. EPA has completed decisions for 11,262 of these products and still must complete decisions for 10,860 products. EPA expects to complete product reregistration in 2014.

FY 2009 Progress and Goals

During FY 2009, EPA completed 1,769 product reregistration decisions, significantly exceeding its goal of 1,275 decisions. EPA's goal is to complete 1,500 product reregistration decisions in FY 2010.

REDs with Product Reregistration Decisions Completed

As of the end of FY 2009, EPA has completed product reregistration decisions for 214 REDs (out of a total of 384 REDs). These 214 REDs include 27 of the 31 organophosphates (OPs).

Process Improvements

During the past several years, as RED production work decreased, EPA has placed increased emphasis on post-RED work and product reregistration. As a result of an external review of the product reregistration process that was completed in 2006 – 2007, EPA has initiated a number of efforts to improve the timeliness and overall productivity of the product reregistration program.

Historical Product Reregistration Decisions

	FY 02	FY 03	FY 04	FY 05	FY 06	FY 07	FY 08	FY 09
Products reregistered	77	53	78	104	169	529	679	603
Products amended	51	40	35	63	40	80	205	292
Products cancelled	186	213	14	342	297	370	309	869
Products suspended	0	5	0	0	0	0	3	5
Total	**314**	**311**	**127**	**509**	**506**	**979**	**1,196**	**1,769**

These efforts include, for example, use of a streamlined batching process, which has reduced the number of product-specific acute toxicity studies required; streamlining the Agency's internal work flow processes; and incorporating quantitative goals into performance standards for managers and staff. As a result of these efforts, the number of product reregistration decisions completed each year has increased significantly.

During FY 2009, EPA completed 1,769 product reregistration decisions, exceeding its goal of 1,275 decisions by almost 40%. The Agency completed almost 15 times as many product reregistration decisions in FY 2009 as it did five years earlier. To strive for continuous improvement, the Agency is again increasing its product reregistration goal for FY 2010.

EPA anticipates that a Memorandum of Understanding for Work-sharing on Product Reregistration developed in FY 2009 and signed in October 2009 between two divisions in EPA's Office of Pesticide Programs (OPP) will enable the Agency to further increase its productivity in FY 2010 and beyond. This MOU establishes the parameters of a work-sharing agreement for product reregistration between OPP's Antimicrobials Division and Pesticide Re-Evaluation Division (formerly the Special Review and Reregistration Division). The two divisions established the agreement to support meeting and exceeding EPA's product reregistration goals. EPA expects that through implementation of the agreement, the product reregistration program will be completed more quickly and risk mitigation measures implemented earlier thereby achieving important human health and environmental protection goals sooner.

STATUS OF REGISTRATION REVIEW

Overall Status

EPA is continuing to meet all registration review targets consistent with overall program objectives. Out of a universe of over 700 registration review cases, including over 1,100 pesticide active ingredients, over 140 cases are past the Preliminary Work Plan stage, and Final Work Plans have been completed for 108 cases. These totals include cases that were scheduled, but which were not required to go through registration review because there are no active domestic registrations.

FY 2009 Accomplishments

During FY 2009, the pace of *registration review* continued to accelerate. Seventy (70) new dockets were opened and 61 Final Work Plans completed.

Registration Review Progress FY 2007 – FY 2009

Fiscal Year	Dockets Opened	Final Work Plans Completed
FY 2007	25	13
FY 2008	46	34
FY 2009	70	61
Total	**141**	**108**

Fiscal year 2009 marked the point at which the Agency concluded ramping up the registration review process and began opening new cases at the pace that must be maintained for the next eight years in order to finish the initial 15-year cycle on schedule in 2022. FY 2009 also featured the first preliminary risk assessments for conventional chemicals under the registration review program. The preliminary risk assessments for clomazone and fomesafen were published and received public comments. The preliminary risk assessments were also provided to the Fish and Wildlife Service and National Marine Fisheries Services (the Services). EPA and the Services are collaborating to improve the process of preparing pesticide risk assessments for consultation under the Endangered Species Act, using clomazone as a pilot case.

Schedule for FY 2010 and Beyond

EPA plans to issue an updated *schedule* for the registration review program in 2010. The new schedule will provide the timeline for opening dockets for the next four years of the program, from FY 2010 to 2013, and will include information on dockets that opened in FY 2007 through FY 2009. The schedule reflects EPA's plan to open about 70 new dockets each year through 2017. This keeps the Agency on track to complete the first 15-year cycle of registration review by October 1, 2022, for all pesticides registered as of October 1, 2007. EPA plans to continue to update the registration review schedule at least annually.

Registration review cases for which Final Work Plans have been developed are currently preceding through the Data Call-In process toward the acquisition of data needed to produce risk assessments consistent with current science, policies, and regulatory requirements. The Agency anticipates publishing additional preliminary risk assessments for public comment, including one for urea sulfate, which was published during the first quarter of FY 2010.

Process Improvements in Registration Review

Section 33(k)(2)(D) requires that the Agency report its recommendations for process improvements in the handling of registration review under Section 3(g) and for streamlining the registration review process. Under the auspices of the PRIA Process Improvement Workgroup, stakeholders have met with Agency staff to identify process improvement opportunities in registration review in a process similar to the one used to identify improvement priorities for the registration process. Recommendations are anticipated in 2010.

In April 2009, EPA formed the Registration Review Listed Species Workgroup within the pesticides program, which is tasked with developing a nationwide risk assessment process to evaluate the impact of pesticide registration on endangered and threatened (listed) species and their critical habitats. This assessment process should be in compliance with the statutory and procedural requirements of section 7 of the Endangered Species Act (ESA) as well as FIFRA. This Workgroup is collaborating with the Services by working on a pilot assessment, focusing on developing a format and content acceptable to the Services. The Workgroup is also focusing on guidance development to allow deadlines to be met and to comply with the statutory mandate for Registration Review. The goal is an efficient and streamlined process that produces assessments consistent with the established agreements between the Agency and the Services (the "Overview Document"), the ESA, FIFRA, and implementing regulations and to efficiently use the Agency's and the Services' resources.

Agency senior staff is defining roles in the evolving registration review process. During FY 2010, the process from preliminary risk assessments through the public comment period and endangered species consultation will be refined based on initial feedback from the clomazone and fomesafen cases. The goal is to adapt the process to provide the public an opportunity to comment on the preliminary risk assessments and then provide a refined consultation package to the Services, while still maintaining a flow of work through the system to allow the Agency to complete registration reviews in a timely manner.

Other Activities

Use of Outside Reviewers

The Agency continued its work-sharing efforts with Canada's Pest Management Regulatory Agency (PMRA), the Australian Pesticides and Veterinary Medicines Authority (APVMA), and the European Union (EU). In global and joint reviews, EPA makes its own registration decision while sharing the study reviews and the risk assessment work and harmonizing its regulatory decisions with other national authorities. One new conventional active ingredient was registered in FY 2009 after a global review and four others were in review. Five new biopesticide active ingredient joint PMRA/EPA reviews were pending at the end of FY 2009 and five more are anticipated in the future.

In FY 2009, PMRA and EPA also implemented a work-share process for minor uses for those chemicals/crops that can not be completed as a joint review. One minor use action was completed as part of the NAFTA work-share program. No joint reviews were completed, but three were pending at the end of FY 2009. Ten additional active ingredient minor use chemicals are expected to be evaluated under the NAFTA joint review program in 2009-2010 and six are expected to be work-share projects.

Table I. PRIA Funded Pesticide Safety Education and Worker Protection Activities in FY 2009

Recipient & Mechanism	Activity and Accomplishments	FY 2009 PRIA Funds
U.S. Department of Agriculture – interagency agreement to pass funds to state cooperative extension services	PRIA funds provide partial support for state level pesticide applicator safety training (classroom, manuals, on line media) to develop competency for existing and potential certified pesticide applicators in using restricted use pesticides safely. The training focuses on a population of applicators (approximately 1,100,000 commercial and private applicators) who can suffer high exposure and risk themselves, or subject others to high exposure and risk, if not trained to competency standards that help to ensure safe pesticide applications. Through an interagency agreement with USDA, funds were transferred to state cooperative extension programs. The funds are distributed by formula based on the numbers of certified applicators reported by the states. The funding formula was revised in 2008 to more accurately reflect the actual workload related to certifying and recertifying Commercial and Private Applicators in each state. With $1.1 million in appropriated funds and $500,000 in PRIA funds, the national total to help support this activity was $1,600,000. The additional funds were allocated to state cooperative extension services by formula based on the number and type of certified applicators reported by the state regulatory agencies. The PRIA funding provides every state extension program with predictable additional resources to support their programs and help ensure that pesticide applicators receive adequate training to competently use restricted-use pesticides.	$500,000
Medical University of South -Carolina cooperative agreement	PRIA funds were used in 2008 to enter into a 2-year cooperative agreement with the Medical University of South Carolina to develop a 6th Edition of the *Recognition and Management of Pesticide Poisoning*. Additional funding was provided in FY 2009. The 5th edition was produced in 1999. The manual, in the English or Spanish	$40,040

Table I. (Continued)

Recipient & Mechanism	Activity and Accomplishments	FY 2009 PRIA Funds
	version, is used internationally to provide health professionals with information on the health hazards of pesticides currently in use, and consensus recommendations for the management of pesticide poisonings and injuries. The fifth edition of this manual is a renowned quick-reference guide to help clinicians identify and properly treat patients with suspected pesticide-related illnesses. The fifth edition needs to be updated to reflect the most current consensus on the appropriate management of pesticide poisonings. The new manual will include a new chapter dedicated to Pyrethroids. The manual will also include more discussion on the health effects of long-term exposure to pesticides. Accomplishments in 2009 include: • Literature review completed for several chapters including neonicotinoids, pyrethroids, fipronil, cancer effects, chlorophenoxy herbicides, Pentachlorophenol, birth defects • New literature review: insect repellents, organophosphates, Paraquat/diquat • List of pesticides and their cancer classifications Drafts of new chapters: • Neonicotinoids section for other insecticides chapter. Status: completed • Fipronil section for other insecticides. Status: completed • Cancer effects. Status: in progress Revisions of old chapters: • Pyrethroids (formerly found in “Other Insecticides”). Status: in progress. • Chlorophenoxy herbicides. Status: completed	

Table I. (Continued)

Recipient & Mechanism	Activity and Accomplishments	FY 2009 PRIA Funds
	• Pentachlorophenol/nitrocresolic herbicides. Status: completed. • Paraquat/diquat. Status: in progress • Insect repellents. Status: near completion • Organophosphates. Status: near completion	
U.S. Department of Labor - interagency agreement	PRIA funds were used to fund the development and analysis of the data in, and for focused reports from, the National Agricultural Workers Survey (NAWS), which contains the most comprehensive demographic information on agricultural workers. Specific questions were developed on worker exposure to assist with worker regulatory development and risk assessments in FY 2008. Questions were pilot tested and revised in summer 2009 and will be included as part of the next extensive NAWS survey cycle. Support for the development of key pesticide worker safety survey questions for the National Agricultural Workers Survey and for focused reports from the survey were continued with remaining funds from the 2008 PRIA funds.	(continued use of 2008 PRIA funds)
National Institutes of Occupational Safety and Health - interagency agreement	PRIA funds support the goal of the NIOSH Sentinel Event Notification System for Occupational Risk (SENSOR) pesticides project to carry out acute pesticide-related illness and injury surveillance activities in the 12 participating states: California, Iowa, Michigan, New York, North Carolina, Texas, Washington, Arizona, Louisiana, Florida, New Mexico, and Oregon. The participating states use consistent criteria for ranking incidents (possible, probable, actual confirmed) and reporting. The confirmed incidents are tracked to medical resolution. The resulting data could be used by the Agency as indicators of the magnitude of incidents; signals of trends	$9,960

Table I. (Continued)

Recipient & Mechanism	Activity and Accomplishments	FY 2009 PRIA Funds
	in exposures; insights into correctable causes of incidents; and data for the risk assessment and registration review processes. Through an interagency agreement with NIOSH, PRIA funds from EPA continue to support pesticide incident surveillance activities in Iowa, Texas, and Washington State. Since the inception of the current interagency agreement in 2005, PRIA funds have supported the expansion of the surveillance program from 9 participating states to 12 participating states. State surveillance work includes the identification, classification and documentation of pesticide poisoning cases. The states periodically perform in-depth investigations of pesticide-related events, and develop interventions aimed at particular industries or pesticide hazards. NIOSH supports these surveillance activities by providing funds and technical support to state health departments. Data submitted annually by each SENSOR state program is aggregated to produce a national database consisting of acute pesticide-related illness and injury cases. The SENSOR-Pesticides program is valuable for providing depth of detail surrounding reported incidents, as well as for the timely identification of emerging trends involving occupational pesticide exposure. This surveillance program is unique in its potential to inform the Agency's Pesticide Program. This quarter, EPA requested that NIOSH provide incident reports for 11 EPA- registered active ingredients under regulatory review. An analysis of this type of incident data is useful in informing OPP risk assessments.	
Association of Farmworker Opportunity	Support is provided to a variety of the national affiliates of AFOP for pesticide worker safety training, education and outreach for farmworkers and farmworker families. This work is targeted at increasing protections for communities	

Table I. (Continued)

Recipient & Mechanism	Activity and Accomplishments	FY 2009 PRIA Funds
Programs (AFOP) - cooperative agreement	with environmental justice issues. These communities have: • a potential for high pesticide exposure, high risk • low literacy, non-English speakers, low income • high mobility • children at risk from take home exposure PRIA funds were used for a cooperative agreement with the Association of Farmworker Opportunity Programs (AFOP) for the following: • Project HOPE (Health and Outreach with Pesticide Education). In a train-the-trainer effort, AFOP has trained over 150 farmworker community outreach workers in 22 sites around the country on how to conduct pesticide worker protection safety education for farmworkers. These outreach workers are, for farmers, a main source of free pesticide worker safety training. • Project SAFE (Saving American Farmworkers Everywhere). Through an EPA/AFOP/AmeriCorps Program, AFOP is training AmeriCorps members to become pesticide safety educators. AmeriCorps members work in AFOP affiliate sites to conduct hands-on, interactive pesticide safety education for farmworkers, farmworker families and other members of the agricultural community in 15 sites across the country. After AmeriCorps members, often grown children of migrant farmworkers, complete one or two years of public service, conducting worker protection pesticide safety trainings, they receive a scholarship award from AmeriCorps to continue or complete their education. • Project LEAF (Limiting Exposures Around Families). In response to research demonstrating higher levels of pesticide in farmworker children and the effectiveness of simple mitigation measures, AFOP developed a program to	$385,000

Table I. (Continued)

Recipient & Mechanism	Activity and Accomplishments	FY 2009 PRIA Funds
	prevent take-home pesticide exposure to farmworker children. • Spanish Radio Campaign to Protection Farmworker Children. AFOP worked with Hispanic Communications Network to create and air a radio campaign on preventing pesticide exposure to farmworker children. A national radio campaign aimed at farmworker parents on how to protect their children was aired in the peak of the 2009 growing season on 245 Spanish language radio stations that reach over 14 million listeners. Additional programs are being developed with new messages aimed at protecting farmworker children to reinforce AFOP's. These will be aired starting in spring, 2010. • Development of a Building Bridges Program. AFOP is partnering with Farmworker Justice to develop a project to assess pesticide safety training programs and other pesticide worker protection programs in Florida. The project will work with key farmworker, grower and state stakeholder groups to develop creative, model pesticide safety education programs to be used by farmers or agricultural service organizations to deliver pesticide worker safety training, as required by federal regulation. • Students Action with Farmworkers (SAF). AFOP worked with SAF (Duke University) to train 75 interns on how to conduct interactive pesticide safety education for farmworkers and farmworker families.	
Pacific Northwest Agricultural Safety and Health	As part of the EPA's health care providers' initiative, support is provided to the University of Washington's Pacific Northwest Ag Safety and Health Center to develop projects to improve the training of health care providers in the recognition,	$132,500

Table I. (Continued)

Recipient & Mechanism	Activity and Accomplishments	FY 2009 PRIA Funds
Center (PNASH) - cooperative agreement	diagnosis, treatment, and prevention of pesticide poisoning among those who work with pesticides, through the development of curricula in several medical, nursing, and public health schools. Generally nurses and physician's assistants may be the first and only health care provider that pesticide workers see. The health care providers' initiative continues to raise awareness among clinicians and nurses nationwide on the recognition and appropriate treatment of patients with symptoms of a pesticide-related illness. PNASH continues to work to incorporate guidance for assessing pesticide exposure into the medical and nursing school curricula. Accomplishments in FY 2009 include the following: **Module Development:** The Problem Based Learning (PBL) case module targets health care workers in a rural clinical setting, but could be adapted to different audiences. It contains instructional information and resources (readings, literature references, online resources, graphic materials, an instructor's guide, and an evaluation process). The module is available to both the UW campus nursing faculty and students. It is being incorporated into the UW Public Health website and will be integrated into the Public Health Medicine site. **Curricula Insertion Points:** Training modules inserted into the following courses Seattle Pacific University Registered Nurse Undergraduate program: • Home Assessment – Environmental Home Assessment Tool • Hospital Survey: Clinical Site Harmful Substances Evaluation University of Washington - Physicians Assistant Program Problem Based Learning – A Pesticide Suicide Attempt	

Table I. (Continued)

Recipient & Mechanism	Activity and Accomplishments	FY 2009 PRIA Funds
	• Oral Herbicide (Diquat) Exposure • A Pyrethroid Inhalation from a Crop Duster • Occupational and Environmental Health History University of Washington – Family Nurse Practitioner Program • Assessing Pesticide Exposure Risk – A Farmworker's Toddler • Occupational and Environmental Health History University of Washington – Nursing • Paresthesias – Is it Diabetes? • Farmworker Housing and Exposure – A Community Case Study Washington State University – Nursing • Organophosphate Exposures – Implications for Nursing Practice • Children's Home Exposure to Pesticides **Website Development and Testing:** • Website structure internally approved at the spring quarterly meeting • The informational components of the website and the learning modules are now in place • The core organizational concepts are flexible and intuitive • PNASH met with Migrant Clinicians Network in order to harmonize materials between the two projects to allow PNASH to put lectures and presentation materials on their network • PNASH conducted a webinar demonstration of the website for MCN and EPA staff	

Table I. (Continued)

Recipient & Mechanism	Activity and Accomplishments	FY 2009 PRIA Funds
Migrant Clinicians Network - cooperative agreement	As part of EPA's National Strategies for Healthcare Providers: Pesticide Initiative, funding through a five year cooperative agreement, begun in 2006, is provided to Migrant Clinicians Network to improve the recognition and management ofpesticide poisoning by healthcare providers in the practice setting. The agreement was formulated in recognition of the need for increased provider competency in environmental and occupational health, in order to support agricultural worker health and safety. For the target populations to be served by the migrant clinicians' network, nurses and physician's assistants may be the first and only health care provider that pesticide workers see. The primary goals of the agreement are to provide resources and training to practicing clinicians, and to develop a model program to incorporate environmental and occupational health (EOH) into the primary care setting. In the fourth year of the agreement, emphasis continued to be placed on a variety of training and presentation activities, resource development, dissemination of pesticide-related health information, and the recruitment of additional health care centers to participate in the program. Throughout Year 4, MCN trained 3 additional health centers to participate in the program, recruited an additional site in Puerto Rico for Year 5, and provided support to ongoing clinical programs. MCN has so far recruited and partnered with 7 clinics to participate in expanded guidance for clinicians to help prevent pesticide related illness. The two additional clinics recruited for Year 5 will bring the total number of programs to 9. MCN continues to stress to participating clinics the importance of incorporating patient occupational exposure histories into the standard clinic procedures in order to identify potential health concerns.	$132,500

Table I. (Continued)

Recipient & Mechanism	Activity and Accomplishments	FY 2009 PRIA Funds
	Through the past year MCN has also provided extensive local, regional and national training sessions on pesticide related issues, reaching over 400 health care providers through 13 clinical sessions, highlighting pesticide exposure management for farmworkers. Additionally, MCN provided 5 general trainings on the issues facing migrant and seasonal farmworkers including overviews of educational materials to reduce risk (153 attendees) and 2 intensive EOH specific sessions (33 attendees). Additionally, MCN has continued to reach its goals of establishing extensive networks linking providers with EOH specialists, and broadened recognition of the health care provider initiative through organization of strategic meetings and presentations at key conferences. Through trainings and other partnerships, MCN distributed a total of 12,560 pesticide-related educational materials and clinical resources for providers and 8,500 patient education materials. 10,800 copies of MCN's bimonthly publication, containing an EOH/pesticide section, were distributed over the fiscal year. MCN continues to maintain its website as a key resource for clinicians seeking information on farmworkers and pesticide exposure, and recorded 20,863 downloads of pesticide related materials. The influence and reach of MCN has increased awareness of the potential health impacts of pesticides among providers, through training and distribution of materials such as EPA's Recognition and Management of Pesticide Poisonings manual.	
Abt, Associates - contract	Abt Associates, through an EPA contract, is conducting economic cost/benefit analyses and other required regulatory analyses that are necessary to support the proposed amendments to the agricultural worker protection regulation and the pesticide applicator certification rule. Because of requirements under Executive	$300,000

Table I. (Continued)

Recipient & Mechanism	Activity and Accomplishments	FY 2009 PRIA Funds
	Orders, Paperwork Reduction Act, and the Regulatory Flexibility Act, these analyses are required to quantifying cost/benefits of the impacts of proposed regulatory changes. In the past year, Abt provided analyses that supported the Small Business Regulatory Enforcement Fairness Act (SBREFA) panel analysis and report and developed new methodology to assess the benefits to be realized from the regulations' amendments. Their new methodology will be used in this analysis. Planned work includes aggregating the line item costs into total societal costs, performing sensitivity analyses and addressing the various regulatory analyses required by executive order and the Regulatory Fairness Act (RFA).	
	2009 Total	**$1,500,000**

Table II. Partnership Grants – Funding and Accomplishments FY 2008 – Partnership Grants

(approximately $970K in grants ($750K in PRIA2 fees and additional appropriated funds) to fund five projects FY 2008 projects run from October 2008 to September 2010)

Recipient	Project Title and Accomplishments	Funding
California Department of Pesticide Regulation (Sacramento, CA):	**"Reducing Volatile Organic Compound Emissions from Pesticide Use in Nuts and Tree Fruit Orchards in California's San Joaquin Valley."** Funds are being used to reduce both pesticides in surface water runoff and volatile organic compound (VOC) emissions from almond, peach, and walnut orchards in California. Funding supported: creation of a multi-agency team of project partners, project team meetings, demonstration of a web-based "VOC Calculator" to potential end users, development of a new	$159,494

Table II. (Continued)

Recipient	Project Title and Accomplishments	Funding
	Conservation Management Practices guide (CMP), and training for farmers on how to use the CMP in conjunction with Year Round IPM plans to reduce pesticides in water runoff and VOC emissions.	
IPM Institute of North America (Madison, WI)	**"High-level IPM in All U.S. Schools by 2015."** PRIA funds are being used to establish and verify adoption of integrated pest management (IPM) in all public kindergartens through high schools in the United States by 2015. This comprehensive approach promotes use of more than presently implemented IPM tools by teaching IPM managers about pest biology, inspection and monitoring for both pests and pest-conducive conditions, and prevention through education, sanitation, and maintenance techniques. Funds have supported the implementation of two new measures (i.e., cockroach allergen levels and student absenteeism) to evaluate the effectiveness of the pest management practices. Funding was used to establish a national network of professionals in five states for promoting proven IPM tools and methods. This goal was exceeded with the addition of nine partner states to the network, four more than originally expected.	$250,000
University of Florida, College of Agriculture and Life Sciences (Gainesville, FL)	**"Reduced Pesticide Use for *Bermisia tabaci* and Greenhouse Whiteflies (GHWF) on Greenhouse Tomato using Protected Culture, IPM Techniques, Parasitic Wasps, and Papaya Banker Plants."** PRIA funds are being used to promote adoption of a biological pest management system that reduces continuous use of pesticides in greenhouses. Papaya banker plants serve as a home base for parasitic wasps. The wasps feed on whiteflies that are greenhouse pests to tomato plants. The funds support demonstrating an IPM approach for controlling both whiteflies and the spread of viral diseases by placing papaya banker plants in tomato greenhouses at four (cooperator owned) demonstration sites in Florida.	$246,418

Table II. (Continued)

Recipient	Project Title and Accomplishments	Funding
Michigan State University (East Lansing, MI)	**"Increasing Adoption of Reduced-Risk Pest Management Practices in Midwest Blueberries to Prepare for FQPA Implementation."** PRIA funds are being used to prepare the Great Lakes' blueberry industry for the phase-out of broad- spectrum pesticides, including azinphos-methyl, by increasing the adoption of reduced-risk alternatives as well as IPM methods. Accomplishments include determining the rain-fastness of nine alternative pesticides in a laboratory; establishing three demonstration field sites with high pest pressure and then field testing three pest control programs that replace Guthion and pyrethroid based insecticides. Preliminary results of the field tests show comparable or better control with reduced-risk alternative pesticides. Two workshops were presented to groups of 50 and 75 attendees (including growers, crop consultants, and industry representatives) on IPM approaches. Throughout the 2009 growing season, 21 weekly newsletters were sent to over 450 recipients. Additional field demonstrations and training will be conducted in a second year of this project.	$91,508
Central Coast Vineyard Team (CCVT) (Paso Robles, CA):	**"Reducing Pesticide Risk through the Adoption of Integrated Farming Practices in Central Coast Vineyards and Marketing Certified Sustainable Products."** PRIA funding is furthering adoption and implementation of IPM practices through CCVT's grower self assessment and the "Sustainability in Practice Vineyard Certification Program (SIP)." Funds also support in-field research to demonstrate effective alternatives to pesticides currently used in vineyards by installing four grower cooperator research stations with Argentine ant bait stations and by working with a grower cooperator to implement an IPM bait-station strategy for the control of mealybugs. CCVT is delivering outreach programs to educate and guide growers on the use of integrated farming systems using "Certified Sustainable Standards," and to educate the public and trade on the environmental and economic benefits of products that are"Certified Sustainable." Funds were used to conduct two workshops on pest	$225,000

Table II. (Continued)

Recipient	Project Title and Accomplishments	Funding
	identification attended by 100 Spanish speaking participants and to conduct training in support of the SIP. Twenty new vineyards applied to participate in the SIP certification program in 2009.	
FY 2009 – Partnership Grants (approximately $950,000 in grants awarded ($750K in PRIA 2 Fees with additional appropriated funds) to fund four projects FY 2009 projects run from October 2009 to September 2011)		
University of California (Berkeley, CA)	**"Implementing reduced risk alternatives for management of codling moth in walnuts"** Funding for this new project supports demonstrating the effective use of a biopesticide (pheromone) currently used for apples to replace azinphos methyl and other pesticides commonly used in growing walnuts. A reduction in the use of these pesticides also addresses OPP's concern for water quality and runoff issues and resistance management in relation to controlling codling moth.	$249,687
University of Wisconsin (Madison, WI)	**"Expanding and Improving the Use of IPM in Midwest Fruit Production"** PRIA funds will be used to support adoption of IPM practices for apples, cherries, and grapes in Wisconsin and other local area states to address water quality and runoff issues. Funds will be used to train new IPM coaches on the use of low-cost modifications to airblast sprayers to help growers in Wisconsin, Minnesota, Illinois, and Iowa use pesticides more efficiently and reduce drift and water runoff. This training will be comprehensive and include the use of reduced risk pesticides as well as biopesticides and pheromone technologies.	$202,027
Baltimore City Health Department (Baltimore, MD)	**"Safe Pest Management for Health (SPMH): An initiative to reduce community pesticide use, increase integrated pest management (IPM), and improve environmental health in Baltimore through public and private partnerships"** PRIA funds will be used to address multiple IPM approaches to improve human health by controlling pests in residences, schools, day care facilities, and homeless service centers in Baltimore, Maryland. Funds will support developing IPM training and site plans for target	

Table II. (Continued)

Recipient	Project Title and Accomplishments	Funding
	sites; developing and administering the nation's first IPM subsidization program for low-income families; coordinating pest control with six partner organizations; and implementing an educational IPM program for Baltimore's Latino community. PRIA funds will also be leveraged by an ongoing separately funded weatherization program to further incorporate IPM into urban structures. This project is projected to train 5,450 persons (including, residents and city staff) and anticipates up to 75% reduction in pesticide uses. Project findings will be disseminated to Baltimore city and national stakeholders through a regional IPM summit, workshops, conferences, journal articles, community presentations, partnership meetings, and web-based materials.	$250,000
The Pennsylvania State University (University Park, PA)	**"Collaborative Design & Delivery of a Unified Training Platform for IPM in Buildings."** PRIA funds will be applied to increasing IPM in urban structures through a pilot training program. Education modules will be developed for "IPM in Buildings" to be first used in the Philadelphia area and later disseminated nationally via an internet-based training program. The modules will address IPM in diverse building types and management systems and developed for service providers and their clients. The pilot program is expected to train 80 owners, 500 health outreach professionals, and 400 occupants.	$249,770

Table III. Number of PRIA Actions Completed in FY 2007, 2008 and 2009

Key to the table

R - Conventional Pesticides, A - Antimicrobial Pesticides, B – Biopesticides, EUP - Experimental Use Permit
PIP - Plant-Incorporated Protectants, SAP - FIFRA Scientific Advisory Panel
SCLP - Straight Chain Lepidopteran Pheromones

PRIA Category	Description of Category	FY 2007		FY 2008		FY 2009	
		Number Completed PRIA Decisions	Average Decision Time in Days	Number Completed PRIA Decisions	Average Decision Time in Days	Number Completed PRIA Decisions	Average Decision Time in Days
R01	New Active Ingredient, Food Use	17	648	34	824	15	648
R010	New Active Ingredient, Food Use					6	570
R02	New Active Ingredient, Food Use, Reduced Risk	10	738	12	446		
R03	New Active Ingredient, Food Use, Experimental Use Permit (EUP) submitted simultaneously with application for registration	1	634				
R04	New Active Ingredient, Food Use, EUP with temporary tolerance, submitted before application for registration	3	195				
R05	New Active Ingredient, Food use submitted after an EUP			12	175		

Table III. (Continued)

PRIA Category	Description of Category	FY 2007		FY 2008		FY 2009	
		Number Completed PRIA Decisions	Average Decision Time in Days	Number Completed PRIA Decisions	Average Decision Time in Days	Number Completed PRIA Decisions	Average Decision Time in Days
R06	New Active Ingredient, Non-food use, outdoor	7	864	1	541	1	753
R060	New Active Ingredient, Non-food use, outdoor					1	245
R07	New Active Ingredient, Non-food use, outdoor, Reduced Risk	0		1	530		
R08	New Active Ingredient, Non-food use, outdoor, EUP request submitted simultaneously with application for registration	2	379				
R09	New Active Ingredient, Non-food use, outdoor, EUP submitted before application for registration	2	205	2	74		
R11	New Active Ingredient, Non-food use, indoor	4	832				
R124	Conditional Ruling on Preapplication Study Waivers; applicant- initiated			1	100	15	153
R130	First food use; indoor; food/food handling			3	325		

Table III. (Continued)

PRIA Category	Description of Category	FY 2007		FY 2008		FY 2009	
		Number Completed PRIA Decisions	Average Decision Time in Days	Number Completed PRIA Decisions	Average Decision Time in Days	Number Completed PRIA Decisions	Average Decision Time in Days
R14	New Use, Additional food use, indoor Food/Food handling	4	715	2	627		
R15	New Use, First Food Use		1456	7	776	9	642
R16	New Use, First Food Use, Reduced Risk					3	555
R17	New Use, Each Additional New Food Use	153	646	186	575	21	714
R170	New Use, Additional food use			2	272	71	440
R18	New Use, Each Additional New Food Use, Reduced Risk	7	865	9	636	2	381
R180	New Use, Additional food use; reduced risk					12	361
R19	New Use, Additional New Food Uses, Bundled, 6 or more	36	691	73	491	11	661
R190	New Use, Additional food uses; 6 or more submitted in one application					23	419
R20	New Use, Additional New Food Uses, Bundled, 6 or more, Reduced Risk	0		3	1274		

Table III. (Continued)

PRIA Category	Description of Category	FY 2007		FY 2008		FY 2009	
		Number Completed PRIA Decisions	Average Decision Time in Days	Number Completed PRIA Decisions	Average Decision Time in Days	Number Completed PRIA Decisions	Average Decision Time in Days
R200	New Use, Additional food uses; 6 or more submitted in one application; reduced risk					14	336
R21	New food use, With EUP and temporary tolerance			1	360		
R220	New Use, Additional food use; EUP; crop destruct basis; no credit toward new use registration			1	96	3	108
R23	New use, Non-food, outdoor	23	632	15	447	3	402
R230	New Use, Additional use; non-food; outdoor			1	285	14	372
R24	New use, Non-food, outdoor, Reduced Risk	7	538	6	403		
R240	New Use, Additional use; non-food; outdoor; reduced risk					5	331
R25	New use, Non-food, outdoor with EUP (no credit toward new use registration)	2	205	1	180		

Table III. (Continued)

PRIA Category	Description of Category	FY 2007		FY 2008		FY 2009	
		Number Completed PRIA Decisions	Average Decision Time in Days	Number Completed PRIA Decisions	Average Decision Time in Days	Number Completed PRIA Decisions	Average Decision Time in Days
R250	New Use, Additional use; non-food; outdoor; EUP; no credit toward new use registration					1	182
R26	New Use, Non-food, indoor	13	507	1	361		
R260	New use; non-food; indoor					2	352
R272	Review of Study Protocol; applicant-initiated; excludes DART, pre-registration conferences, Rapid Response review, DNT protocol review, protocols needing HSRB review			12	82	48	68
R274	New Uses, Additional uses; seed treatment only; 6 or more submitted in one application; limited uptake into Raw Agricultural Commodities; includes crops with established tolerances (e.g., for soil or foliar application); includes food and/or non-food uses					1	359
R28	Import tolerance, New Active Ingredient or first food use	2	688				

Table III. (Continued)

PRIA Category	Description of Category	FY 2007		FY 2008		FY 2009	
		Number Completed PRIA Decisions	Average Decision Time in Days	Number Completed PRIA Decisions	Average Decision Time in Days	Number Completed PRIA Decisions	Average Decision Time in Days
R280	Establish import tolerance; new active ingredient or first food use					1	637
R29	Import tolerance, Additional new food use	2	597	3	1000	1	1000
R290	Establish import tolerance; additional food use					1	432
R292	Amend an established tolerance (e.g., decrease or increase); domestic or import; applicant-initiated			7	221	16	329
R293	Establish tolerance(s) for inadvertent residues in one crop; applicant-initiated					3	317
R30	New Product, Me-Too, Fast Track	301	73	103	75		
R300	New product; identical or subst-antially similar in composition and use to a registered product; no data review or only product chemistry data; cite-all data cita-tion, or selective data citation where applicant owns all requi-red data, or applicant submits specific authorization letter from data			169	74	239	76

Table III. (Continued)

		FY 2007		FY 2008		FY 2009	
PRIA Category	**Description of Category**	**Number Completed PRIA Decisions**	**Average Decision Time in Days**	**Number Completed PRIA Decisions**	**Average Decision Time in Days**	**Number Completed PRIA Decisions**	**Average Decision Time in Days**
	owner. Category also includes 100% re-package of regist-ered end-use or manufacturing-use product that requires no data submission nor data matrix.						
R301	New product; identical or subst-antially similar in composition and use to a registered product; registered source of active ingre-dient; selective data citation only for data on product chemi-stry and/or acute toxicity and/or public health pest efficacy, where applicant does not own all required data and does not have a specific authorization letter from data owner.			21	116	43	122
R31	New Product, Non-Fast Track (includes review of product chemistry, acute toxicity, public health pest efficacy)	337	183	193	204	2	873
R310	New end-use or manufacturing-use product; requires review of data package within RD;			100	166	236	194

Table III. (Continued)

PRIA Category	Description of Category	FY 2007		FY 2008		FY 2009	
		Number Completed PRIA Decisions	Average Decision Time in Days	Number Completed PRIA Decisions	Average Decision Time in Days	Number Completed PRIA Decisions	Average Decision Time in Days
	includes reviews and/or waivers of data for only: • product chemistry and/or • acute toxicity and/or • public health pest efficacy						
R313	New product; requires amend-ment to existing inert tolerance exemption (e.g., adding post-harvest use); applicant-initiated					4	416
R32	New Product, Non-Fast Track, new physical form (excludes selective citations)	8	356	14	349	3	513
R320	New product; new physical form; requires data review in science divisions			1	141	9	346
R33	New manufacturing-use product, Old Active Ingredient, Selective Citation	20	472	18	461	1	551
R330	New manufacturing-use prod-uct; registered active ingredient; selective data citation					3	386
R331	New Product; repack of identi-cal registered end-use product as a					1	77

Table III. (Continued)

PRIA Category	Description of Category	FY 2007		FY 2008		FY 2009	
		Number Completed PRIA Decisions	Average Decision Time in Days	Number Completed PRIA Decisions	Average Decision Time in Days	Number Completed PRIA Decisions	Average Decision Time in Days
	manufacturing-use product; same registrant uses only						
R34	Amendment, Non-Fast Track (includes changes to precauti-onary label statements, source changes to an unregistered source)	179	111	64	119	1	421
R340	Amendment requiring data review within RD (e.g., changes to precautionary label statem-ents, or source changes to an unregistered source of active ingredient)			95	100	200	111
R35	Amendment, Non-Fast track (changes to REI, PPE, PHI, rate and number of applications, add aerial application, modify GW/SW advisory statement)	45	380	48	255		
R350	Amendment requiring data review in science divisions (e.g., changes to REI, or PPE, or PHI, or use rate, or number of applications; or add aerial application; or modify GW/SW advisory statement)			15	184	59	215

Table III. (Continued)

		FY 2007		FY 2008		FY 2009	
PRIA Category	**Description of Category**	**Number Completed PRIA Decisions**	**Average Decision Time in Days**	**Number Completed PRIA Decisions**	**Average Decision Time in Days**	**Number Completed PRIA Decisions**	**Average Decision Time in Days**
R37	Cancer Reassessment, applicant initiated	3	785	6	536		
A41	New Active Ingredient, Non-food use, outdoor, other uses	6	879	3	125	2	
A42	New Active Ingredient, Non-food use, indoor, FIFRA sec. 2(mm) uses	2	644	2	998	2	920
A44	New Use, First food use, with exemption	1	41	3	739	2	682
A46	New Food Use, with exemption	6	497	3	470	2	492
A460	Additional Food use; establish tolerance exemption					6	199
A470	New Food use, Additional food use; establish tolerance					1	436
A48	New use, Non-food, outdoor FIFRA sec. 2(mm) uses	1	261	3	262		
A480	New use, Additional use; non-food; outdoor; FIFRA §2(mm) uses			1	20	1	391
A49	New use, Non-Food, outdoor, other uses	2	436	2	460		

Table III. (Continued)

PRIA Category	Description of Category	FY 2007		FY 2008		FY 2009	
		Number Completed PRIA Decisions	Average Decision Time in Days	Number Completed PRIA Decisions	Average Decision Time in Days	Number Completed PRIA Decisions	Average Decision Time in Days
A490	New use, Additional use; non-food; outdoor; uses other than FIFRA §2(mm)					3	454
A50	New use, Non-food, indoor FIFRA sec. 2(mm) uses	7	253	11	412	1	1002
A500	New use, Additional use; non-food; indoor; FIFRA §2(mm) uses			3	228	22	263
A520	Experimental Use Permit application					2	181
A521	Review of public health efficacy study protocol within AD; per AD Internal Guidance for the Efficacy Protocol Review Pro-cess; applicant-initiated; Tier 1			3	146	1	342
A522	Review of public health efficacy study protocol outside AD by members of AD Efficacy Protocol Review Expert Panel; applicant-initiated; Tier 2					1	234
A53	New Product, Me-too, Fast Track	80	108	25	102		

Table III. (Continued)

PRIA Category	Description of Category	FY 2007		FY 2008		FY 2009	
		Number Completed PRIA Decisions	Average Decision Time in Days	Number Completed PRIA Decisions	Average Decision Time in Days	Number Completed PRIA Decisions	Average Decision Time in Days
A530	New product; identical or subs-tantially similar in composition and use to a registered product; no data review or only product chemistry data; cite-all data citation, or selective data citation where applicant owns all required data, or applicant submits specific authorization letter from data owner. Category also includes 100% re-package of registered end-use or manufacturing-use product that requires no data submission nor data matrix.			45	73	49	95
A531	New product; identical or subst-antially similar in composition and use to a registered product; registered source of active ingre-dient; selective data citation only for data on product chemi-stry and/or acute toxicity and/or public health pest efficacy, where applicant does not own all			4	90	12	151

Table III. (Continued)

PRIA Category	Description of Category	FY 2007		FY 2008		FY 2009	
		Number Completed PRIA Decisions	Average Decision Time in Days	Number Completed PRIA Decisions	Average Decision Time in Days	Number Completed PRIA Decisions	Average Decision Time in Days
	required data and does not have a specific authorization letter from data owner.						
A532	New product; identical or subst-antially similar in composition and use to a registered product; registered active ingredient; unregistered source of active ingredient; cite-all data citation except for product chemistry; product chemistry data submitted			12	66	19	156
A54	New Product, Non-Fast Track, FIFRA sec. 2 (mm) uses	75	178	44	224	3	446
A540	New end use product; FIFRA §2(mm) uses only			25	110	70	139
A55	New Product, Non-Fast Track, other uses	10	254	5	222	1	615
A550	New end-use product; uses other than FIFRA §2(mm); non- FQPA product			2	172	4	180
A56	New Manufacturing use product, old active ingredient, selective citation	5	418	5	470		

Table III. (Continued)

PRIA Category	Description of Category	FY 2007		FY 2008		FY 2009	
		Number Completed PRIA Decisions	Average Decision Time in Days	Number Completed PRIA Decisions	Average Decision Time in Days	Number Completed PRIA Decisions	Average Decision Time in Days
A560	New manufacturing-use prod-uct; registered active ingredient; selective data citation			2	176	3	349
A57	Amendments, Non-Fast Track	113	129	55	115	1	454
A570	Label amendment requiring data submission			78	105	136	123
B59	New Active Ingredient, Food Use, with exemption, Microbial/ Biochemical	9	654	3	737		
B590	New active ingredient; food use; establish tolerance exemption, Microbial/Biochemical					3	487
B60	New Active Ingredient, Non-food use, Microbial/Biochemical	6	485	6	980	6	732
B600	New active ingredient; non-food use, Microbial/Biochemical					7	385
B61	EUP, Food Use with temporary tolerance exemption, Microbial/-Biochemical	5	251	2	349		
B610	Food use; EUP; establish temporary tolerance exemption, Microbial/Biochemical					6	286

Table III. (Continued)

		FY 2007		FY 2008		FY 2009	
PRIA Category	Description of Category	Number Completed PRIA Decisions	Average Decision Time in Days	Number Completed PRIA Decisions	Average Decision Time in Days	Number Completed PRIA Decisions	Number Completed PRIA Decisions
B62	EUP, Non-food use, Microbial/-Biochemical	1	196				
B620	Non-food use; Experimental Use Permit application					4	127
B621	Extend or amend EUP, Microbial/Biochemical			3	62	3	101
B63	New Use, First Food Use, with tolerance exemption Microbial/ Biochemical,	2	356	8	459	4	541
B630	First food use; establish tolerance exemption, Microbial/ Biochemical					2	313
B631	Amend established tolerance exemption, Microbial/ Biochemical			1	270	3	242
B65	New Use, Non-Food, Microbial/ Biochemical	2	337				
B650	New use; Non-Food, Microbial/ Biochemical					2	239
B66	New Product, Me-Too, Fast Track, Microbial/biochemical	14	69	4	94		

Table III. (Continued)

PRIA Category	Description of Category	FY 2007		FY 2008		FY 2009	
		Number Completed PRIA Decisions	Average Decision Time in Days	Number Completed PRIA Decisions	Average Decision Time in Days	Number Completed PRIA Decisions	Average Decision Time in Days
B660	New product; identical or substantially similar in composition and use to a registered product; no data review or only product chemistry data; cite-all data citation, or selective data citation where applicant owns all required data, or applicant submits specific authorization letter from data owner. Category also includes 100% re-package of registered end-use or manufacturing-use product that requires no data submission nor data matrix. Microbial/ biochemical			9	79	6	73
B67	New Product, Non-Fast Track, Microbial/Biochemical	35	184	23	282	1	895
B670	New product; registered source of active ingredient; all Tier I data for product chemistry, toxicology, non-target organisms, and product performance must be addressed with product specific			7	161	9	282

Table III. (Continued)

PRIA Category	Description of Category	FY 2007		FY 2008		FY 2009	
		Number Completed PRIA Decisions	Average Decision Time in Days	Number Completed PRIA Decisions	Average Decision Time in Days	Number Completed PRIA Decisions	Average Decision Time in Days
	data or with request for data waivers supported by scientific rationales, Microbial/Biochemical						
B672	New product; non-food use or food use having established tol-erance or tolerance exemption; unregistered source of active ingredient; no data compensate-ion issues; all Tier I data requi-rements for product chemistry, toxicology, non-target organi-sms, and product performance must be addressed with product-specific data or with request for data waivers supported by scientific rationales, Microbial/ Biochemical					12	280
B68	Amendment, Non-Fast Track, Microbial/Biochemical	28	122	7	115		
B680	Label amendment requiring data submission, Microbial/ Biochemical			4	195	9	129
B681	Label amendment; unregistered source of active ingredient;					3	244

Table III. (Continued)

		FY 2007		FY 2008		FY 2009	
PRIA Category	**Description of Category**	**Number Completed PRIA Decisions**	**Average Decision Time in Days**	**Number Completed PRIA Decisions**	**Average Decision Time in Days**	**Number Completed PRIA Decisions**	**Average Decision Time in Days**
	supporting data require scientific review						
B682	Protocol Review; applicant-initiated; excludes time for HSRB review (pre-application)					1	89
B69	Straight Chain Lepidopteran Pheromones (SCLP), New Active Ingredient, Food or Non-Food Use	1	235				
B690	SCLP, New active ingredient; food or non-food use			1	180	2	231
B700	SCLP, Experimental Use Permit application; new active ingredient or new use					1	134
B71	SCLP, New Product, Me-Too, Fast Track	5	75				
B710	SCLP, New product; identical or substantially similar in composition and use to a registered product; no data review or only product chemistry data; cite-all data citation, or selective data citation where applicant owns all			3	93	5	94

Table III. (Continued)

PRIA Category	Description of Category	FY 2007		FY 2008		FY 2009	
		Number Completed PRIA Decisions	Average Decision Time in Days	Number Completed PRIA Decisions	Average Decision Time in Days	Number Completed PRIA Decisions	Average Decision Time in Days
	required data, or applicant submits specific authorization letter from data owner. Category also includes 100% re-package of registered end-use or manufacturing-use product that requires no data submission nor data matrix						
B72	SCLP, New Product Non-Fast Track	6	209	2	194		
B720	SCLP, New product; registered source of active ingredient; all Tier I data for product chemi-stry, toxicology, non-target org-anisms, and product performa-nce must be addressed with product specific data or with request for data waivers suppo-rted by scientific rationales					7	143
B730	SCLP, Label amendment requiring data submission					3	72
B74	Plant-Incorporated Protectants (PIP), EUP, Non Food/feed or	1	318				

Table III. (Continued)

PRIA Category	Description of Category	FY 2007		FY 2008		FY 2009	
		Number Completed PRIA Decisions	Average Decision Time in Days	Number Completed PRIA Decisions	Average Decision Time in Days	Number Completed PRIA Decisions	Average Decision Time in Days
	crop destruct No Scientific Advisory Panel (SAP) review						
B740	Plant-Incorporated Protectants (PIP), EUP; registered active ingredient; non-food/feed or crop destruct basis; no Scientific Advisory Panel (SAP) review required					2	135
B75	PIP, EUP, with Temporary Tolerance or Exemption, No SAP review	2	268	1	269		
B77	PIP, EUP, New Active Ingredient, Set Temporary Tolerance or Exemption, SAP			2	517		
B772	Amend or extend EUP; minor changes to experimental design; established temporary tolerance or tolerance exemption is unaffected			1	96	3	76
B81	PIP, Register New Active Ingredient, Temporary Tolerance or Exemption Exists, SAP	3	635	3	539	3	587

Table III. (Continued)

PRIA Category	Description of Category	FY 2007		FY 2008		FY 2009	
		Number Completed PRIA Decisions	Average Decision Time in Days	Number Completed PRIA Decisions	Average Decision Time in Days	Number Completed PRIA Decisions	Average Decision Time in Days
B810	New active ingredient; establish permanent tolerance or tolerance exemption based on temporary tolerance or tolerance exemption; SAP review required			1	538		
B86	PIP, EUP, Food Use, Amendment	1	147	5	208		
B880	PIP, New product; no SAP review required					2	382
B90	PIP, Amendment, Non-Fast Track	2	179				
B900	Amendment (except #B890); No SAP review required; (e.g., new IRM requirements that are applicant initiated; or amending a conditional registration to exte-nd the registration expiration date with additional data submitted)			2	176	14	176
B904	Import tolerance or tolerance exemption; processed commodities/food only					1	103
	Total	**1620**		**1677**		**1570**	

Table IV. Number of PRIA Applications Pending at the End of Fiscal Year (FY 2007 Through FY 2009)

Key to the table
R - Conventional Pesticides
A - Antimicrobial Pesticides
B - Biopesticides
EUP - Experimental Use Permit
PIP - Plant-Incorporated Protectants
SAP - FIFRA Scientific Advisory Panel
SCLP - Straight Chain Lepidopteran Pheromones

Progress in Meeting Decision Times – Number of PRIA Actions Pending at End of Fiscal Year				
PRIA Category	**Description of Category**	**Number of PRIA Decisions Pending at the End of Fiscal Year**		
		2007	**2008**	**2009**
R01	New Active Ingredient, Food Use	48	24	5
R010	New Active Ingredient, Food Use		14	29
R02	New Active Ingredient, Food Use, Reduced Risk	18	6	
R020	New Active Ingredient, Food use; reduced risk			4
R05	New Active Ingredient, Food use submitted after an EUP	17	1	
R06	New Active Ingredient, Non-food use, outdoor	6	5	3
R060	New Active Ingredient, Non-food use, outdoor		11	45
R07	New Active Ingredient, Non-food use, outdoor, reduced risk	1		
R09	New Active Ingredient, Non-food use, outdoor, EUP submitted before application for registration	2		
R10	New Active Ingredient, Non-food use, outdoor, submitted after EUP	3	3	3
R11	New Active Ingredient, Non-food use, indoor	2		

Table IV. (Continued)

PRIA Category	Description of Category	Number of PRIA Decisions Pending at the End of Fiscal Year		
		2007	2008	2009
R110	New Active Ingredient, Non-food use; indoor			3
R120	New Active Ingredient, Non-food use; indoor; reduced risk		2	2
R123	New Active Ingredient, Seed treatment only; includes non-food and food uses; limited uptake into Raw Agricultural Commodities		2	2
R124	Conditional Ruling on Preapplication Study Waivers; applicant-initiated		11	2
R13	New Use, First food use, indoor food/food handling	2	2	2
R14	New Use, Additional food use, indoor Food/Food handling	5	3	
R15	New Use, First Food Use	18	11	2
R16	New Use, First Food Use, Reduced Risk	3	3	
R17	New Use, Each Additional New Food Use	255	67	31
R170	New Use, Additional Food Use		112	135
R18	New Use, Each Additional New Food Use, Reduced Risk	11	2	
R180	New Use, Additional food use; reduced risk		13	17
R19	New Use, Additional New Food Uses, Bundled, 6 or more	93	31	20
R190	New Use, Additional food uses; 6 or more submitted in one application		34	60
R20	New Use, Additional New Food Uses, Bundled, 6 or more, Reduced Risk	4		
R200	New Use, Additional food uses; 6 or more submitted in one application; reduced risk		11	19
R21	New food use, With EUP and temporary tolerance	2	1	

Table IV. (Continued)

PRIA Category	Description of Category	Number of PRIA Decisions Pending at the End of Fiscal Year		
		2007	2008	2009
R210	New Use, Additional food use; EUP; establish temporary tolerance; no credit toward new use registration		2	
R220	New Use, Additional food use; EUP; crop destruct basis; no credit toward new use registration		2	
R23	New use, Non-food, outdoor	24	7	3
R230	New Use, Additional use; non-food; outdoor		25	23
R24	New Use, Non-food, outdoor, Reduced Risk	6		
R240	New Use, Additional use; non-food; outdoor; reduced risk		4	
R25	New use, Non-food, outdoor with EUP (no credit toward new use registration)	1		
R250	New Use, Additional use; non-food; outdoor; EUP; no credit toward new use registration		1	5
R26	New Use, Non-food, indoor	4	1	
R260	New use; non-food; indoor		3	6
R270	New use; non-food; indoor; reduced risk			1
R272	Review of Study Protocol; applicant-initiated; excludes DART, pre-registration conferences, Rapid Response review, DNT protocol review, protocols needing HSRB review		7	9
R273	Additional use; seed treatment; limited uptake into Raw Agricultural Commodities; includes crops with established tolerances (e.g., for soil or foliar application); includes food or non-food uses			10
R274	New Uses, Additional uses; seed treatment only; 6 or more submitted in one application; limited uptake into Raw Agricultural Commodities; includes crops		3	2

Table IV. (Continued)

PRIA Category	Description of Category	Number of PRIA Decisions Pending at the End of Fiscal Year		
		2007	2008	2009
	with established tolerances (e.g., for soil or foliar application); includes food and/or non-food uses			
R28	Import tolerance, New Active Ingredient or first food use	1	1	1
R280	Establish import tolerance; new active ingredient or first food use		2	3
R29	Import tolerance, Additional new food use	9	5	3
R290	Establish import tolerance; additional food use		2	9
R292	Amend an established tolerance (e.g., decrease or increase); domestic or import; applicant-initiated		22	27
R293	Establish tolerance(s) for inadvertent residues in one crop; applicant-initiated		3	
R295	Establish tolerance(s) for residues in one rotational crop in response to a specific rotational crop application; applicant-initiated		1	4
R30	New Product, Me-Too, Fast Track	85		
R300	New product; identical or substantially similar in composition and use to a registered product; no data review or only product chemistry data; cite-all data citation, or selective data citation where applicant owns all required data, or applicant submits specific authorization letter from data owner. Category also includes 100% re-package of registered end-use or manufacturing-use product that requires no data submission nor data matrix.		51	73

Table IV. (Continued)

PRIA Category	Description of Category	Number of PRIA Decisions Pending at the End of Fiscal Year		
		2007	2008	2009
R301	New product; identical or substantially similar in composition and use to a registered product; registered source of active ingredient; selective data citation only for data on product chemistry and/or acute toxicity and/or public health pest efficacy, where applicant does not own all required data and does not have a specific authorization letter from data owner.		13	15
R31	New Product, Non-Fast Track (includes review of product chemistry, acute toxicity, public health pest efficacy)	189	4	
R310	New end-use or manufacturing-use product; requires review of data package within RD; includes reviews and/or waivers of data for only: • product chemistry and/or • acute toxicity and/or • public health pest efficacy		141	112
R311	New product; requires approval of new food-use inert; applicant-initiated; excludes approval of safeners			2
R313	New product; requires amendment to existing inert tolerance exemption (e.g., adding post-harvest use); applicant-initiated		6	1
R32	New Product, Non Fast Track, new physical form (excludes selective citations)	17	5	1
R320	New product; new physical form; requires data review in science divisions		9	15
R33	New manufacturing-use product, Old Active Ingredient, Selective Citation	21	1	
R330	New manufacturing-use product; registered active ingredient; selective data citation		3	8

Table IV. (Continued)

PRIA Category	Description of Category	Number of PRIA Decisions Pending at the End of Fiscal Year		
		2007	2008	2009
R34	Amendment, Non-fast Track (includes changes to precautionary label statements, source changes to an unregistered source)	58	1	
R340	Amendment requiring data review within RD (e.g., changes to precautionary label statements, or source changes to an unregistered source of active ingredient)		63	55
R35	Amendment, Non-fast track (changes to REI, PPE, PHI, rate and number of applications, add aerial application, modify GW/SW advisory statement)	48	2	2
R350	Amendment requiring data review in science divisions (e.g., changes to REI, or PPE, or PHI, or use rate, or number of applications; or add aerial application; or modify GW/SW advisory statement)		63	76
R37	Cancer Reassessment, applicant initiated	6	1	1
R370	Cancer reassessment; applicant-initiated			1
A380	New Active Ingredient, Food use; establish tolerance exemption		1	1
A400	Non-food use; outdoor; FIFRA section (2mm) uses			1
A41	New Active Ingredient, Non-food use, outdoor, other uses	5	2	2
A42	New Active Ingredient, Non-food use, indoor, FIFRA sec. 2(mm) uses	5	3	1
A420	Non-food use; indoor; FIFRA section 2(mm) uses			4
A44	New Use, First food use, with exemption	5	2	
A440	New Use, First food use; establish tolerance exemption		2	2
A46	New Food Use, with exemption	4	1	

Table IV. (Continued)

PRIA Category	Description of Category	Number of PRIA Decisions Pending at the End of Fiscal Year		
		2007	2008	2009
A460	New Food Use, Additional food use; establish tolerance exemption		3	7
A470	New Food use, Additional food use; establish tolerance		1	
A48	New use, Non-food, outdoor FIFRA sec. 2(mm) uses	4		
A480	New use, Additional use; non-food; outdoor; FIFRA §2(mm) uses		2	4
A49	New use, Non-Food, outdoor, other uses	2	1	
A490	New use, Additional use; non-food; outdoor; uses other than FIFRA §2(mm)		3	1
A50	New use, Non-Food, indoor FIFRA sec. 2(mm) uses	10	1	
A500	New use, Additional use; non-food; indoor; FIFRA §2(mm) uses		27	18
A510	Additional use; non-food; indoor; uses other than FIFRA section 2(mm)			2
A520	Experimental Use Permit application		1	3
A521	Review of public health efficacy study protocol within AD; per AD Internal Guidance for the Efficacy Protocol Review Process; applicant-initiated; Tier 1		2	2
A522	Review of public health efficacy study protocol outside AD by members of AD Efficacy Protocol Review Expert Panel; applicant-initiated; Tier 2		3	3
A53	New Product, Me-too, Fast Track	22		
A530	New product; identical or substantially similar in composition and use to a registered product; no data review or only product chemistry data; cite-all data citation, or selective data citation where applicant owns all required data, or applicant submits specific authorization letter from data owner. Category also		15	18

Table IV. (Continued)

PRIA Category	Description of Category	Number of PRIA Decisions Pending at the End of Fiscal Year		
		2007	2008	2009
	includes 100% re-package of registered end-use or manufacturing-use product that requires no data submission nor data matrix.			
531	New product; identical or substantially similar in composition and use to a registered product; registered source of active ingredient; selective data citation only for data on product chemistry and/or acute toxicity and/or public health pest efficacy, where applicant does not own all required data and does not have a specific authorization letter from data owner.		5	2
A532	New product; identical or substantially similar in composition and use to a registered product; registered active ingredient; unregistered source of active ingredient; cite-all data citation except for product chemistry; product chemistry data submitted		6	7
A54	New Product, Non-Fast Track, FIFRA sec. 2 (mm) uses	41	4	1
A540	New end use product; FIFRA §2(mm) uses only		32	48
A55	New Product, Non-Fast Track, other uses	4	1	
A550	New end-use product; uses other than FIFRA §2(mm); non-FQPA product		1	
A56	New Manufacturing use product, old active ingredient, selective citation	5		
A560	New manufacturing-use product; registered active ingredient; selective data citation		4	6
A57	Amendments, Non-Fast Track	50	1	
A570	Label amendment requiring data submission		55	55

Table IV. (Continued)

PRIA Category	Description of Category	Number of PRIA Decisions Pending at the End of Fiscal Year		
		2007	2008	2009
B59	New Active Ingredient, Food Use, Microbial/Biochemical, with exemption	11	12	8
B590	New active ingredient; food use; establish tolerance exemption, Microbial/ Biochemical,		19	31
B60	New Active Ingredient, Non-food use, Microbial/Biochemical	14	10	1
B600	New active ingredient; non-food use, Microbial/Biochemical,		9	12
B61	EUP, Food Use with temporary tolerance exemption, Microbial/Biochemical	2		
B610	Food use; EUP; establish temporary tolerance exemption, Microbial/ Biochemical,		2	
B63	New Use, First Food Use, Microbial/Biochemical, with exemption	11	7	3
B630	First food use; establish tolerance exemption, Microbial/Biochemical,		1	5
B631	Amend established tolerance exemption, Microbial/Biochemical		4	1
B650	New use; Non-Food, Microbial/ Biochemical			1
B66	New Product, Me-Too, Fast Track, Microbial/biochemical	4		
B660	New product; identical or substantially similar in composition and use to a registered product; no data review or only product chemistry data; cite-all data citation, or selective data citation where applicant owns all required data, or applicant submits specific authorization letter from data owner. Category also includes 100% re-package of registered end-use or manufacturing-use product that requires no data submission nor data		2	4

Table IV. (Continued)

PRIA Category	Description of Category	Number of PRIA Decisions Pending at the End of Fiscal Year		
		2007	2008	2009
	matrix. Microbial/biochemical			
B67	New Product, Non-Fast Track, Microbial/Biochemical	27	1	
B670	New product; registered source of active ingredient; all Tier I data for product chemistry, toxicology, non- target organisms, and product performance must be addressed with product specific data or with request for data waivers supported by scientific rationales, Microbial/Biochemical		5	10
B671	New product; food use; unregistered source of active ingredient; requires amendment of established tolerance or tolerance exemption; all Tier I data requirements for product chemistry, toxicology, non- target organisms, and product performance must be addressed with product-specific data or with request for data waivers supported by scientific rationales, Microbial/Biochemical		6	6
B672	New product; non-food use or food use having established tolerance or tolerance exemption; unregistered source of active ingredient; no data compensation issues; all Tier I data requirements for product chemistry, toxicology, non-target organisms, and product performance must be addressed with product-specific data or with request for data waivers supported by scientific rationales, Microbial/Biochemical		13	11
B68	Amendment, Non-Fast Track, Microbial/Biochemical	6		
B680	Label amendment requiring data submission, Microbial/Biochemical		6	12

Table IV. (Continued)

PRIA Category	Description of Category	Number of PRIA Decisions Pending at the End of Fiscal Year		
		2007	2008	2009
B681	Label amendment; unregistered source of active ingredient; supporting data require scientific review		2	5
B690	SCLP, New active ingredient; food or non-food use		2	
B710	SCLP, New product; identical or substantially similar in composition and use to a registered product; no data review or only product chemistry data; cite-all data citation, or selective data citation where applicant owns all required data, or applicant submits specific authorization letter from data owner. Category also includes 100% re-package of registered end-use or manufacturing-use product that requires no data submission nor data matrix.		2	
B72	SCLP, New Product Non-Fast Track	2		
B720	SCLP, New product; registered source of active ingredient; all Tier I data for product chemistry, toxicology, non-target organisms, and product performance must be addressed with product specific data or with request for data waivers supported by scientific rationales		1	1
B721	New product; unregistered source of active ingredient			3
B730	SCLP, Label amendment requiring data submission			1
B740	Plant-Incorporated Protectants (PIP), EUP; registered active ingredient; non-food/feed or crop destruct basis; no Scientific Advisory Panel (SAP) review required		1	2
B77	PIP, EUP, New Active Ingredient, Set Temporary Tolerance or Exemption, SAP	1		

Table IV. (Continued)

PRIA Category	Description of Category	Number of PRIA Decisions Pending at the End of Fiscal Year		
		2007	2008	2009
B771	Experimental Use Permit application; new active ingredient; establish temporary tolerance or tolerance exemption; no SAP review required;			2
B80	PIP, Register New Active Ingredient, Temporary Tolerance or Exemption Exists, No SAP	1		
B800	New active ingredient; establish permanent tolerance or tolerance exemption based on temporary tolerance or tolerance exemption; no SAP review required			5
B81	PIP, Register New Active Ingredient, Temporary Tolerance or Exemption Exists, SAP	7	3	
B84	PIP, Register New Active Ingredient, Set Tolerance or Exemption, SAP	2	2	2
B86	PIP, EUP, Food Use, Amendment	3		
B880	PIP, New product; no SAP review required		2	19
B881	PIP, New product; SAP review required		1	1
B900	PIP, Amendment (except #B890); No SAP review required; (e.g., new IRM requirements that are applicant initiated; or amending a conditional registration to extend the registration expiration date with additional data submitted) (2)		13	
B901	PIP, Amendment (except #B890); SAP review required		1	1

EPA also continued working with the California Department of Pesticide Regulation (CDPR) to expand capacity to review residue chemistry studies and conduct dietary risk assessments in support of registration decisions. In FY 2009, CDPR reviewed the residue chemistry studies for three active ingredients and a total of 12 representative commodities or crops and completed dietary risk assessments for two active ingredients.

Performance-Based Contracts

Contractors tasked with the review of hazard and exposure data continued to assist the Agency in the selection of endpoints and characterization of hazards for human health and ecological risk assessment. These contractor services enhanced the production of the Agency's risk assessments. The level of contractor support in FY 2009 was approximately the same as in FY 2008 and 2007, and approximately 80% of the Pesticide Program's active contracts or work assignments were performance-based, the same as FY 2008. Performance based contracts tend to be contracts with routine and predictable work assignments. Areas covered by these contracts include information management, records management, on-site computer leasing and support, outreach, and as appropriate, data review and risk assessment.

Appendix A. Decision Review Times for Actions Completed during FY 2009

As required by FIFRA Section 33(k), the following table (an Excel file) provides the decision times for each decision (application) completed during FY 2009. Decisions with a two digit PRIA code are PRIA 1 decisions, while those with a three digit PRIA code are PRIA 2 decisions. Decisions received by the Agency on or after October 1, 2007, are PRIA 2 decisions. Decision times indicated in red with an asterisk are decisions completed before the Agency received payment or a waiver was granted. Completion of a registration action before payment is received typically occurs in situations where a voluntary fee payment has been offered for an application that was pending with the Agency prior to March 23, 2004 (the PRIA effective date), or the Agency anticipates approval of a fee waiver based on past fee waiver approvals during the same maintenance fee cycle.

Mandatory decision time frames depend on the year the application was received. Mandated time frames can be found in the fee schedules published in the Federal Register Notice on March 17, 2004, titled *Pesticides; Fees and Decision Times for Registration Applications* for PRIA 1 actions and for PRIA

2 decisions, and on October 30, 2007, titled *Pesticides; Revised Fee Schedule for Registration Applications.* The Agency's target due date for completing a decision or action under PRIA 1 was based on 30 days in a month. The time frames specified in the Consolidated Appropriations Act of 2004 are in months. In the table, if the PRIA due date was met, while the Agency's target date was not, a date was entered in the column labeled PRIA Due Date. After PRIA 2 was implemented the Agency modified its tracking system to calculate due dates in months. As EPA improves its reporting capabilities, the Agency may update this table, as necessary.

Table of completed actions for FY 2009 (Excel, 276 KB) *(Microsoft Excel Viewer* is needed to view this file.)

CHAPTER SOURCES

The following chapters have been previously published:

Chapter 1 – This is an edited, reformatted and augmented version of Congressional Research Service publication, Report RL31921, dated February 1, 2010.

Chapter 2 – This is an edited, reformatted and augmented version of Congressional Research Service publication, Report RL32884, dated June 15, 2010.

Chapter 3 – This is an edited, reformatted and augmented version of Congressional Research Service publication, Report RL32218, dated June 14, 2010.

Chapter 4 – This is an edited, reformatted and augmented version of Implementing the Pesticide Registration Improvement Act, dated February 26, 2010.

INDEX

D

E

F

G

H

I

J

K

L

M

N

O

P

Q

R

S

T

U

V

W